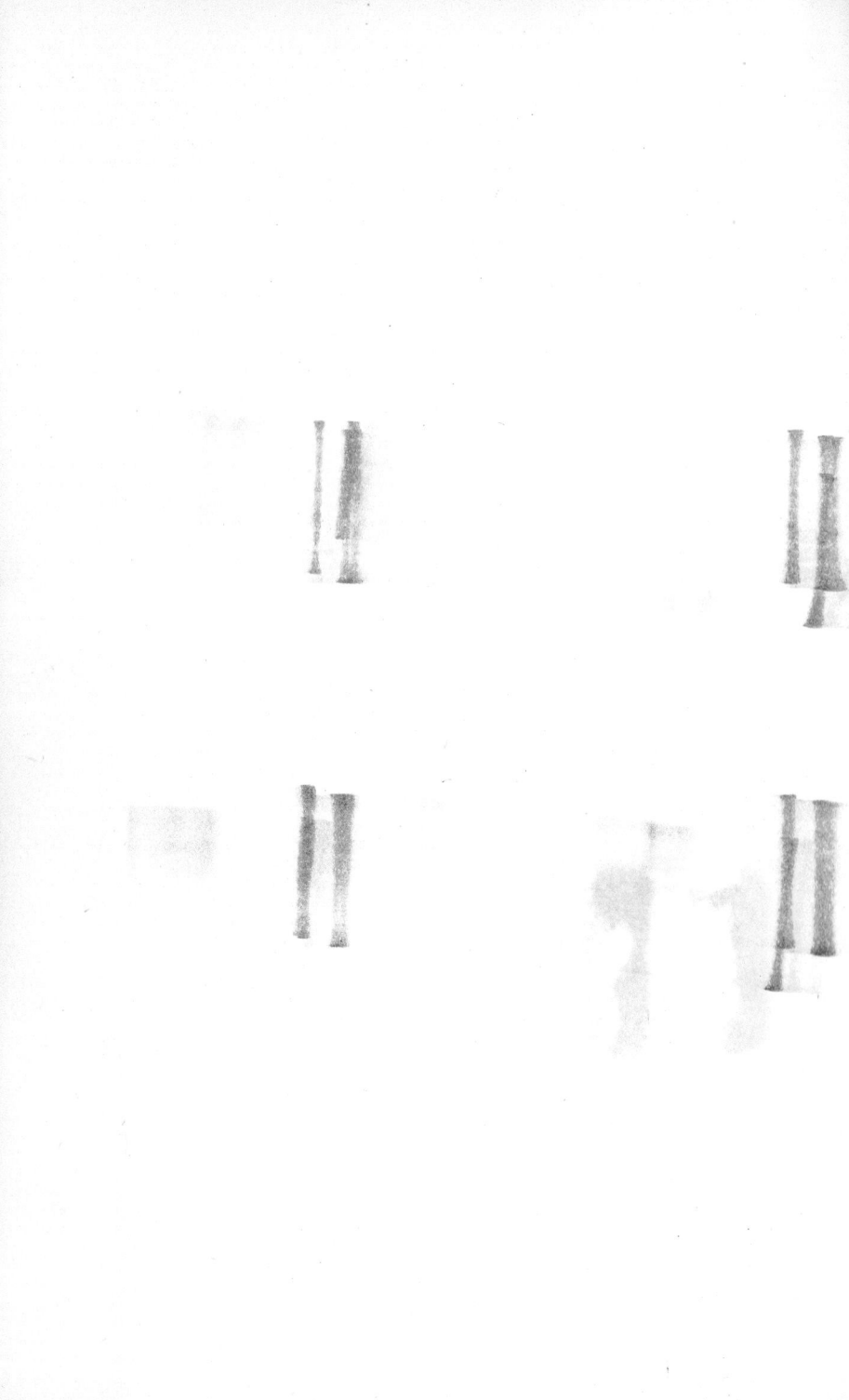

Automation

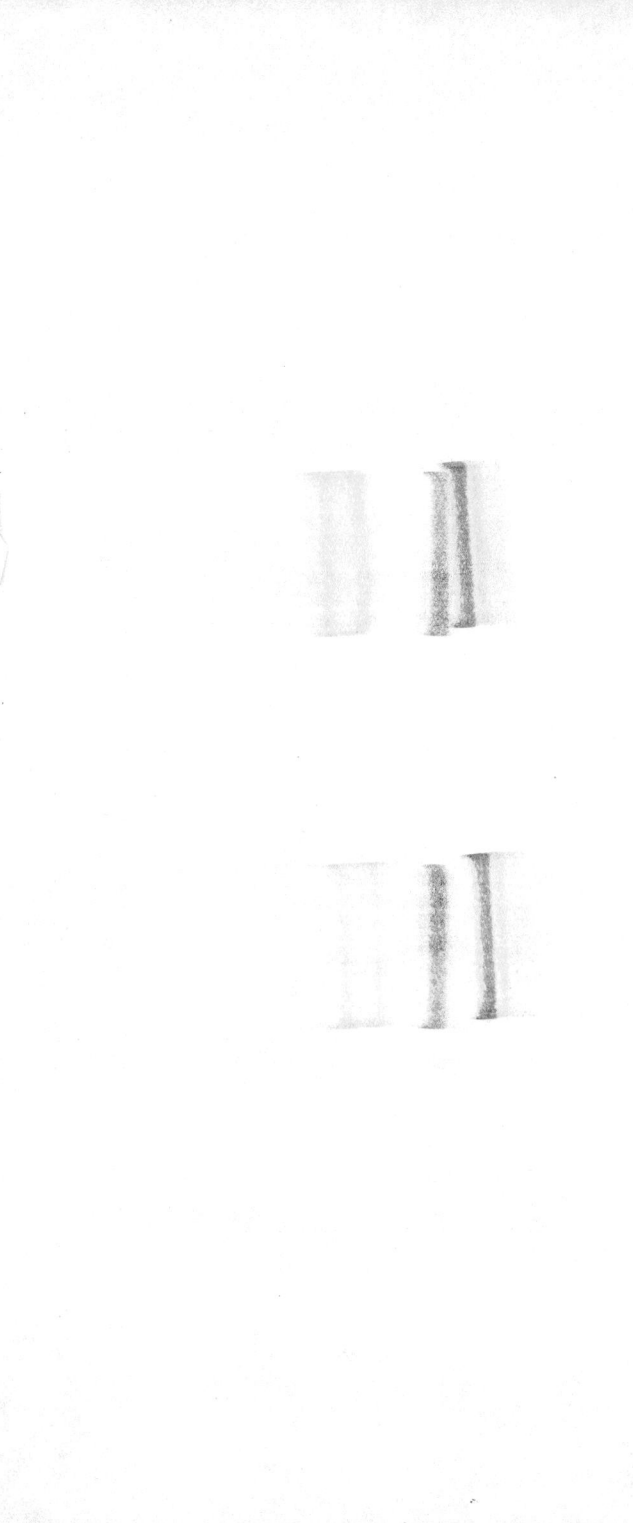

AUTOMATION

JOHN DIEBOLD

amacom
American Management Associations

Library of Congress Cataloging in Publication Data

Diebold, John, 1926-
    Automation.

    Includes index.
    1. Automatic control. I. Title.
TJ213.D49   1983   629.8   83-2763
ISBN 0-8144-5756-8

First Printing

To

BRIGADIER GENERAL GEORGES F. DORIOT

whose own life and accomplishments
are the best lesson to his students

Contents

Editor's Note

WHEN ONE of my associates first suggested to me the possibility of republishing this book in its original form, I had to admit—albeit with some embarrassment—that although I was familiar with the title of John Diebold's first published work I had never actually read it. By way of explanation, I pointed out that I wasn't yet old enough to read when the book was released in 1952. Unfazed by the reminder that an entire generation had grown up since the book's initial publication, he urged me to read it, suggesting that I "might find it interesting."

In fact, I found the book fascinating. Written at a time when *automation* was a new word (which John helped bring into everyday use) and room-size computers performed fewer functions than today's desktop models, the book predicted a world that few of John's contemporaries could envision—a world startlingly like the one in which I and my contemporaries of the postwar baby boom have come of age.

> During the last decade . . . [he wrote at the time] developments in the fields of electronics, communications, and electric network analysis have

made possible the construction of a wide variety of . . . machines capable of automatically performing a sequence of logical operations, similar in many ways to the mental processes of humans. . . . These recent advancements have been of such importance that they constitute the first stages of what coming generations will look upon as a second industrial revolution.

Thus, with the computer industry still in its infancy, John foresaw a time when automation would not only change how we do things but—even more important—change the things we do. In this respect, the second industrial revolution would have an effect on the world very similar to that of its predecessor.

John insisted that one of the most important changes necessitated by this revolution would be in our attitudes. "To harvest all of its [automation's] fruits, we must, in many cases, redesign our products, our processing methods, and our machines. And to redesign them we must *rethink* them." The New York Stock Exchange, he suggested, was a good example of the kind of operation that required rethinking. He considered it "a glaring anachronism," a place where "as in the ancient marketplaces, the traders stand at their posts and offer wares—not stone jugs, but stocks and bonds." It's necessary, he said, to "start by thinking of basic functions," which in the case of the exchange floor is "to provide a means for interchange of information concerning the sale and purchase of stocks and bonds."

"Can this function be performed automatically?" he asked. "Computers are designed primarily to receive, manipulate, and communicate information. They provide a means for eliminating the exchange floor altogether." This was a radical suggestion in its day, but he proceeded to outline a means of automating the entire system that has in large measure been implemented in the intervening years.

The U.S. steel industry—then considered the basic industry of our economy—was used as another example of the need for redesigning product, process, and/or machinery in order to take full advantage of the new technology.

Many automatic control devices are used in steel production, but, with the single exception of continuous casting, steel production is not automatic. Nor is it likely to become automatic, regardless of how many electronic devices we employ, until there is a basic change in the way we make steel.

He went on to suggest a means of automating the steel industry. To our detriment, this method was not adopted in the United States, but it was adopted to a great extent by the Japanese as they rebuilt their factories after the Second World War. The result of this has been that although the Soviet Union and the United States are still the world's largest producers of steel, Japan is now the third largest and is gaining quickly.

Of most immediate interest to managers, however, are John's remarks about the automatic handling of information or, more specifically, what is today called office automation.

It is probably there that the most immediate, widespread, and fruitful application of the new technology will be made. . . . As business grows and becomes more complex, the need for detailed, up-to-the-minute, accurate information increases enormously. Better methods of production control and market analysis, a growing body of government regulations, complex payroll deductions—all these are placing an increasing burden on office procedures.

Computers, he argued, could lead to enormous improvements and economies in the processing of information. As an example of the possible use of computers in office operations, he discussed the retrieval of policy information in the insurance industry.

By the use of magnetic tapes, the storage space for the policy information of a large insurance company can be reduced from ten or fifteen floors of files to 350 or 400 spools of magnetic tape which, with control

gear, would occupy one medium-size room. . . . Under such an arrangement each employee who requires policy information would have on his
desk a dial, similar to a telephone dial, and a small ticker tape printer. . . .
He would dial a code number (which might even be a policy number) and
thereby activate a series of relays; . . . the information on the spool would
be "read" by a decoding device and printed on the ticker tape on the
caller's desk—all within a minute.

The principles of information storage and retrieval John
outlines here are the same as those now employed not only
in the insurance industry but in many other businesses as
well. His prediction fell short only in that he didn't foresee
the desktop terminal in use today or the speed with which
it operates.

As another of the many examples of how computers
could be a boon to business, John wrote:

The computer can answer the many "What would happen if . . . ?"
questions that cannot now be answered. Management can create projected operating data according to various hypotheses. By analyzing these
hypothetical reports of costs, production, and profits, under various sets
of operation conditions, much can be learned about the wisdom of
alternative courses of action.

This is, of course, the process that is referred to today as
computer simulation, which is used in companies all over
the world.

Because computers were not being widely used at the
time of the book's first publication, John's concerns were
mainly with the technical aspects of automation and the
need for rethinking processes. But he also recognized that,
like the first industrial revolution, the second would have a
substantial impact on society, and he placed great importance on the need to direct developments in this area.
Although his predictions about the *extent* of the changes
involved were not entirely correct, he foresaw the *direction*
of these changes quite accurately.

"The most reasonable expectation," he wrote, "is that the

changes accompanying automation will occur at least at the pace of the last two decades [the thirties and forties] and, in all probability, somewhat more rapidly." In fact, of course, the rate of change has accelerated at an almost alarming speed. Even John—let alone any of his less prescient contemporaries—could hardly have foreseen how, in the relatively brief span of thirty years, computers would be produced at a size and price to make them accessible to virtually any middle-class household.

What can we learn, then, from reading *Automation*? It shows us where the second industrial revolution began, how much has already changed, and how much can and must yet be accomplished before humankind realizes the full benefits of this new technology. The control and management of these potential benefits should be one of our major concerns today. John discusses this area at much greater length in the new introduction that follows, but it might be useful to recall here the remarks with which he closed the original edition of this book. I find these thoughts as relevant today as they were in 1952.

Automation . . . will raise very real problems for the human race. . . . But these problems are not altogether new. Just as automation is part of a longer continuum, so too the problems which automation will raise have been with us, in varying forms, for many years. Some of these problems seem to solve themselves, while others require a conscious effort for solution. Many, as is all too evident from the world about us, have not yet been solved. For it is indeed hard to provide a society in which increased material welfare truly benefits man rather than cheapens him. Strong moral leadership and men of good will are sorely needed, as much now as always.

Robert A. Kaplan
Editor-in-Chief
AMACOM Book Division
1983

Looking Ahead

John Diebold, 1983

THE OCCASION of the thirtieth anniversary of the publication of *Automation* has led not only to my rereading my book for the first time since writing it, but to my thinking ahead—to what I would address if I were writing it now for the first time.

I have long believed that any important technological innovation brings about three phases of change. First you mechanize what you did yesterday; second you find that what you do changes; and third you find that, as a result of these changes, the greatest change of all occurs in *society.* When this book was first published in 1952, many important advances in technology were just being made, or just beginning to move into the first phase of mechanizing what had been done before. By now we are into the second and third phases, actually doing different work and coping with the changes that technology has brought about in society.

The three phases of change are hard to recognize while you are living through them, but that is precisely what we must do to capitalize on change. In the first stage we see time and again that success will go to a few companies in each industry that seize market share or increase profits by recognizing how technology can change what they have been doing. However, it inevitably follows that the entire industry moves to a new level of competition or finds itself in new businesses, and what it does changes. Finally, and most important, these changes do not and cannot occur in a vacuum. Society is adjusting to these changes, too. People as individuals are moving into new occupations, they find new work relationships, and their leisure changes. There is increased social mobility, and our rights and freedoms are adjusted to conform to a new reality.

This is not just a national phenomenon. It is worldwide. What is especially fascinating is that the society is also molding the technology to make it more suitable for people. But before we look at the present and future in more detail, let's look back at history for a moment to get a broader perspective.

A HISTORICAL PERSPECTIVE ON TECHNOLOGICAL CHANGE

One of my interests as a student was in what today has the very distinguished title of industrial archeology. Two factories of the eighteenth century, in particular, were at the heart of my initial interest in the area of automation. One factory was designed and built by an itinerant inventor named Oliver Evans who travelled up and down the Delaware Valley in the late 1700s. He built a series of fully automated grain mills, and he had his share of problems. When he tried to get a patent, Thomas Jefferson, who at that time had charge of the patent-issuing process, denied the request. In his letter, Jefferson said the design for Oliver Evans's fully automated grain mill was only a collection of fundamental principles, such as the screw of Archimedes and the inclined plane of Egypt! In fact, it really did have all of the important elements of an automatic factory, including some feedback devices.

The other eighteenth-century factory that influenced me very much as a student was Adam Smith's hypothetical pin factory, which he used in *The Wealth of Nations* (published in 1776) as an example of the way we had come to specialization of labor and very high productivity. This particular notion was absolutely correct, but in working on the question of how best to use automation technology, I was concerned that the Oliver Evans model of complete integration be followed rather than the Adam Smith model, and my feeling was that as mechanization developed in the nineteenth century and in the early part of this century, we followed the pin factory. We began to mechanize the speciality. We began to mechanize around sharpening the point. We began to build machines to attach the head. I am, of course, using this example figuratively because this really was how all of our mechanization developed, around specialized labor.

What this specialization meant is that we were focusing  our attention on discrete actions and on connecting them together in the simplest possible series. However, this type of specialization ignored the complex implications of moving from one specialized activity to the next. We refined the actions, but really did not address their connections into "trans-actions."

One of the things that concerned me very much when I began to work in this field was that suddenly, for the first time, we had the means to really take account of a transaction when it took place, and of its impact throughout the entire enterprise, whether that enterprise was government, an educational institution, or business. Organizationally, however, we were completely tied to Adam Smith's pin factory, and we still are.

I use these examples because I've always found the perspective of history to be very helpful, particularly when trying to get a feeling for what actual consequences will result from some of the developments in technology that are taking place. And the developments get more and more exciting. Some of the new systems are wonderful. The question of what we will do with them is the biggest problem we face as a society.

Again, to give a little perspective of history: I've always been fascinated by the industrial revolution. I think one of the questions it immediately raises, as it did when I was doing my student work, is *why* it was a revolution. Was it the machines? The conclusion I came to was that it wasn't the machines. The steam engine, the cotton gin, the railway, the power loom—all were extraordinary inventions. But the reason they were revolutionary was that they were agents for great *social* change. They were revolutionary because they took people out of the fields and brought them into factories. They gave us mass production and, through mass production, the first society in which wealth was *not* confined to the few. The industrial revolution produced a sense of hurry, a sense of time, a sense of goal that simply didn't exist previously. It changed human society. That is what was revolutionary, not the machines themselves.

Looking at what is happening today, I think the same thing is true. If you had asked Richard Arkwright or James Watt whether they thought they were changing society, they certainly would not have thought so. They were simply concentrating on what they were doing. One of our problems is that we are changing society with much of the new technology. It's very important to be conscious of that fact, and to think much more widely, as many of our leading scientists are doing, of what the social consequences and the human meaning will be.

THE CURRENT ELECTRONICS REVOLUTION

We are now engaged in an electronics revolution, but we have to see this revolution in a long-term perspective to understand its significance. As I noted within this book, "Automation will appear as a distinct phase in industrial progress, but it is nevertheless a part of the long continuum of man's mechanization of his work. The economic and social effects of the new technology should be viewed in this perspective."

It is very easy to get swept up in publicity about revolutionary technological changes. In fact, we are making revolutionary progress in the sense that we are moving ahead and in new directions from a very substantial base. Our current electronics revolution builds on a heritage of technological revolutions that enables us to adjust to change almost unconsciously. This is what enables us to proclaim widely that there is a revolution and then to virtually ignore the complex economic, business, and social issues that we must master before we can say that we have succeeded with our revolution.

Affecting Productivity and the Products Themselves

It seems to me that some of the really interesting developments have not yet started or, if they have, are not quite recognized yet. Either way, they will be very important in the years ahead. The computer industry has developed and grown in the last 30 years to what is today a big industry, but it has already started to shift. It is no longer only a capital industry producing capital goods. It has also become a consumer-product industry, and that is going to

be very significant. Computers are involved in producing not only products, but services, and not only the direct products, but the indirect products—the incorporation of chips into automobiles and into consumer and capital products of all kinds. There are a lot of important consequences from this, because it means that international economic competitiveness in almost all other services and product areas will be determined more and more by computers and the information industry. Consumers will increasingly judge the quality of the goods and services they buy according to the computer support they get with them.

So what we see emerging is a technology that has increasingly embedded itself in our economic infrastructure. The availability of computer resources is becoming as crucial to national economic viability as the availability of energy and raw materials. If the success of the Japanese tells us anything, it should tell us that. The fact that computers are becoming a consumer field, and that they are also going to materially change the productivity of virtually all other businesses and the kinds of products and services they offer, is of the utmost importance in terms of our nation's well-being.

Productivity growth is stimulated by new technology—not only because automation streamlines work, but, more important, because it gives us an opportunity to analyze the jobs we do as individuals, as work groups, and as large systems at the enterprise level. The process of rethinking is truly difficult. We are learning this as we struggle to find ways to measure the productivity of our own actions, especially in the office of today.

It has long been felt that the essence of any definition of productivity was the idea of more efficient operations, defined narrowly. Using this type of paradigm, an entire school of thought examining productivity was brought into being. Now, however, with the increasing penetration of office automation into the workplace, the definition of productivity in its classical sense is being challenged. At the heart of this transformation in thinking lies the realization that office automation not only will change the efficiency with which work is performed, but will change the nature and definition of work itself. (In other words, people are beginning to think about the second phase of change.)

It is useful to consider an analogy with general systems theory. General systems are known to operate using a defined set of criteria and rules of interaction and structure until the feedback throughout the system becomes of such proportions as to cause a fundamental disequilibrium in the system. When this occurs, a "state change" is effected. This is what is happening as we automate information in the office environment.

In its narrowest sense, information productivity can be thought of as a means to achieve profits through improvement in measurable output. However, increasingly we are finding that a completely output-oriented measurement is not adequate for understanding the effects of office automation. If we think of the office environment as a system, certain rules governing the operating efficiency of its units can be derived. Using such rules, we can make progress in identifying criteria for measuring and improving efficiency.

However, just as with a general system, an office will also undergo a fundamental transformation when subjected to a certain amount of stress. When this occurs in the office environment, not only will structural relationships change, but human and conceptual relationships will also change.

At the individual level the changes start to spill over from one person to another and one office to another. But unless we focus our attention on the process, we lose control of what is happening to the system as a whole. And so, no matter whether we can prove changes in productivity or not, it is an essential part of managing to continually strive to analyze productivity.

Right now too much attention is being focused on simplistic cost/benefit equations to improve organizational productivity. Instead, measuring productivity should be a process of trying to analyze the quantity and quality of what we do. It should enable us to better choose our own destiny as we strive to employ new technology in the ways that are best suited to what we intend to do.

Improving Life for Human Beings

The second observation I would make in looking at the years ahead is that the changes occurring in technology involve extraordinary technical steps that will lead to material decreases in costs, and this means that the machine will do more and more complex tasks that will make work easier for human beings. Basically the human/machine interface is getting much easier, and that is what I believe is the real meaning of the technological changes. For example, voice machines now respond to the human voice, using any voice you want; other machines that handle graphics can be communicated with in graphics; and both these machines will soon be easily portable. Technological changes like these mean that we can begin to build systems that are able to handle the flow of information (which is our society's lifeblood), and that are friendly, are easy to use, and can adapt themselves to the human need in this area. These are the kinds of forces that are revolutionizing the entertainment industry and enabling personal computing to be a realistic tool for people in the mainstream of our society. And these are the kinds of changes that let us go faster through a supermarket line or confirm our ticket reservations for us. They mean we can redesign a building or a car or a boat without ever touching pen to paper. That's a very important kind of change, and it's composed of hundreds of innovations.

The intertwining of such technological advances with the other great developments of our times—the biological developments—is inevitable and obvious. For example, biologists have discovered that there are enormous volumes of information encoded within cells. This immediately shows that our machine systems could eventually store far more information in far less space. Obviously it's possible to go very, very much further than most of the people thinking about it realize. *The intertwining of technological and biological developments is going to be formidable.*

Before too long, our use of computer devices in the medical profession is going to lead to very distinct changes in the ways we regard ourselves and the ways we take care of ourselves. The CAT scan is one of the first computer tools to gain wide use in assisting diagnosis. Now all sorts of tests are being automated, including very sophisticated analysis of blood. Teleconferencing and computer-aided instruction are becoming key aspects of medical training. At the same time we are beginning to use artificial-intelligence techniques to aid in diagnosis.

What this means is that we are starting to apply computer tools to understanding ourselves and rethinking the way we ourselves function as human beings as well as biological beings. We are making real strides toward understanding how we function organically, and we are learning much more about our biological ability to handle and store information. We have seen modest progress in terms of biofeedback, and we are starting to understand better our fantastic ability to perceive and retain visual images, how truly complicated it is to distinguish voice messages, how we learn and forget languages, and so on. All these are at the systems level of human beings.

And we are probing much deeper into our physical levels as well. Our ability to create and manipulate genetic structures has come about through the availability of so-phisticated computer analysis tools, but as we learn how to manipulate these complex microscopic codes, we are ap-proaching the development of new ways of storing and processing information in living tissues. We are on the verge of programming seeds to manufacture plant life to fit our needs and designing bacterial organisms to auto-mate our chemical waste disposals.

All this is to say that, far from undermining our appreci-ation for life, automation is enabling us to gain a new appreciation of just how much complexity there is and how much latent potential there is in the biological realm.

Several years ago, Vannevar Bush said we would end up with computers implanted in each of us. We already have chips implanted in heart devices, and we will have a multi-plicity of increasing human involvement in this, but it also shows us the way toward doing a great deal in terms of circuitry.

Changing the Nature of What We Do
The third observation that I have is the fact that, so far, most of the use of automation has been to mechanize work we have done or already are doing. To a certain extent the second and enormously important phase has started, and that is to *change* what we do. Although it is about as hard as it was when I wrote my book to find examples, which was a terrible problem at that point, it is possible nevertheless to find cases showing that the parameters of competition in business are being completely changed as a result of the imaginative use of this technology.

What is happening all around us is that major corporations start encroaching on each other's territory. Entirely new and exciting forms of competition spring up. One salient example of this is, of course, the battle forming now between the titans of telecommunications and computing, but there are other exciting, creative, and less confrontational kinds of crossbreeding taking place. For example, a number of retailers are finding that they can offer extensive financial services that used to be the highly regulated preserve of banks. At the same time, banks can offer financial services at home and sell merchandise over the wires.

A financial service finds retail business opportunities in cable television for home entertainment, and the same kinds of wires are starting to provide internal networks for major corporations. Newspapers and publishing houses are moving into electronic publishing and suddenly finding themselves head to head with giants of the aerospace industry, which have turned their research libraries into on-line information services.

In each of these areas new business opportunities have arisen because technology has enabled major enterprises to reassess their potential and rethink what they can do.

The computer games industry, a single example, is already twice the size of the movie industry in the United States. It is an industry that didn't even exist a few years ago, and it is totally dependent on computers and interactive TV or TV-like screens.

During the past half-dozen years the computer industry has been radically transformed by electronics and communications. Everywhere we see astonishing progress in electronic funds transfer, cash-flow management, electronic tellers, and global networks. Soon banking at home will have moved well beyond the pilot stages. Only a few years ago these services were peripheral to banking activities. Now they mark the competitive edge that one institution has over another.

In retail we see a similar phenomenon, with the spread of bar codes and point-of-sale (POS) terminals. POS technology contributes to effective inventory management and sales strategy. Statistics on the sales flow of products can be used to adjust shelf-space allocations to optimal proportions by product and package, and so improve space utilization and sales appeal. The results of promotional efforts can be quickly determined. Another natural extension is a purchase-order system, which will generate orders as inventory falls to critical levels according to sales flow.

The use of POS terminals allows overnight centralized control of terminals from the corporation's host computer. Customer service is also improved, since there is greater facility for handling charge card and credit sales.

Improved check and credit handling can be a strategic way in which stores with automated sales systems can offer better service to customers and at the same time achieve greater security against loss.

The initial step is being taken toward establishing information links between retail businesses' information systems and those of their suppliers or of manufacturers. The National Retail Merchants Association, which has promoted optical character recognition (OCR) as a suggested standard for sales tags, has also published standards for electronic purchase orders and invoices. These point to savings of time and labor and greater efficiency for both the retail business and the supplier. Electronic exchange of purchase orders, invoices, and stock-on-hand reports eliminates duplication of computer entry work as well as previous manual clerical labor; adjustments become easier and errors are largely eliminated, and the need for telephone and postal communications is reduced. Also, because of improved communications, a shorter lead time is required for filling orders, and this allows both retailer and supplier to reduce their inventories.

So far I have been talking about better computing and communications within companies. Now this all leads to a revolution in terms of electronic exchanges between and among different enterprises, and also to and from the individual consumer. A prime example of this is the way that on-line systems are enabling us to deregulate our transportation industries by using nationwide data bases for air, rail, and road transport to allow optimal routing of goods. All this leads to our increased ability to get products through shorter and more effective delivery cycles—from unprocessed raw materials, through different manufacturing processes, into distribution outlets, and through to users—with very close and efficient tracking. At the same time we can monitor the costs and the movements of capital.

The consequences are astonishing. They have impact on where and how our meat is slaughtered and packaged; they are determining the cost-effectiveness of hydroponics; and they are helping us reduce our needs for energy to move goods physically and store them in proper environments.

Speaking of environments brings me to the rather old concept, but new practice, of automating environments. We are starting to hear more and more about automated buildings and computerized climate controls. It is now becoming cheaper to install controls and wires than to use energy that benefits no one. These same wires and controls can supply computer tools to white collar workers at all levels and run a complex assembly line. Recently we have even seen the introduction of a robot that is operated by the same brand of personal computer that managers use for their own work.

It is in this final area that we are really starting to see significant change. The work that managers do is being radically restructured. Working with words and numbers, or meeting and communicating with people, is where almost all the activities of managers and professionals are being done. Their work used to be supported by numerous underlings and secretaries, but now computers are doing more and more of the filing and sending and formatting and reformatting. They are transforming our business environment because we can execute the work we do much more quickly. Another example is the change going on in the field of finance.

We are totally changing the structure of industries as a result not of mechanizing what we did yesterday, but of doing something quite new—providing services that weren't provided previously. This will be the dominant characteristic of the years ahead: we will provide services and products that simply didn't exist in the past.

A few years ago the stereotype was that developments in automation and computers meant rigid systems: highly centralized, monolithic types of structures that were dehumanizing. Today it's exactly the reverse. Systems are becoming highly decentralized, very flexible, very human, increasingly friendly, and very easy to adapt in any way wanted by the people who are using them. This leads to the ability to unleash human imagination—the most important force we have. And what imagination will lead us to is something we can hardly begin to guess. This is what is really happening in this field.

WILL THE TECHNOLOGICAL REVOLUTION REALLY BENEFIT SOCIETY?

If we take the term *revolution* from the first industrial revolution, we should realize that whatever we do, however badly we mess up or however brilliantly we perform, we are going through an extraordinary social revolution, and the changes in computers and in automation are one of the principal forces bringing about that change. Our task is to make really sure that this change will be for the benefit of mankind and not to its detriment.

Once again, looking back a little bit, the nineteenth century was a very interesting period in American history. It was a period of really massive change. We took relatively crude technology and we did heroic things with it. We threw railroads across the continent; we moved the population off the farms and into factories and cities. We changed the entire face of our country, and the rest of the industrial world followed in this change as well. The tools with which we did it were very simple. We had a strong sense of mission and what we were about. We could do extraordinary things.

Today, by contrast, we have extraordinary technology that is growing more extraordinary every day, yet it has become increasingly difficult to do even the simplest things in our society. I think it's worthwhile looking at why this is true, because it seems to me to be the key to what we will do with the technology and the science that are available to us. I think that any view of the future has to be based on some real analysis of how well we are going to cope with some of these problems as a nation.

Public Policy Problems
To date, the developments in the field of automation technology seem to have raised relatively few discernible public policy problems. Some of us feel that a lot of problems have been raised, but very few people perceive things that way at this point. During the period immediately ahead of us, an increasing array of public policy

problems will be related to technology. What is a branch of a bank? Is a terminal a branch? What should our country's communications policy be, let alone the world's? That question is currently being debated in Washington, perhaps not with the degree of perspective that many of us would like to see, but that is characteristic of virtually all of our public policy considerations.

The process we use to set public policy reflects the highly fragmented structure of our government. We are therefore ill-equipped to deal cohesively with the increasing number of problems that involve many areas at once. One example is antitrust actions. I cannot help feeling that it would be a fascinating irony if what the two Japanese companies have been indicted for—theft of information from IBM—is precisely the information that the European Economic Community is trying to obtain through its antitrust actions. Its proposed remedy in the antitrust action would give the EEC Commission access to precisely the information that the Japanese, according to the indictments, have been responsible for stealing.

The other irony of that situation is that if the EEC prevails, the beneficiary will be the Japanese, because the one thing the European organizations have demonstrated in the last 30 years is an inability to bring to market competitive high-technology products, and the one thing that the Japanese have demonstrated is the ability to do so. But what I am trying to point out are immediate, real, public policy problems, and I believe we're going to find many more of them.

Not only are our problems formidable, but in general they are debated at an entirely incorrect level. Whatever the crisis issues of the moment may be—energy, or inflation, or interest rates—I think those are merely symptoms.

I will give some examples of the kinds of underlying issues that really determine how we can make the maximum use of technological innovation for society's benefit.

Four Fundamental Institutional Problems

One pressing question that confronts us is how we in the United States can maintain our ability to innovate, which is what has put us in the leadership position we enjoy in the technology of automation. Our great strength in innovation comes from various sources. It comes from the mobility of a highly educated population and from the American belief in backing small enterprises, in getting things started. There is a whole mix of things that have made us very dynamic in this regard. But how do we now ensure that the marvelous cornucopia of new technology and inventions will continue to yield the things that our society rests on? Will the Silicon Valley, Route 128, the Hudson Valley, and other parts of the country continue to contribute so enormously to technology?

We have a society in which more and more factors work against taking risk. How do we maintain one in which there is a value system that fosters risk? I think this issue is worth a lot of attention and will determine much of what happens next.

A second problem is one of time horizons. We have a very short time scale in our decision making at both the private and the public level. We tend to use a variety of very sophisticated tools, such as discounted cash flow, which lead to decisions favoring short- rather than long-term investments. This is particularly true in periods of high interest rates. We have a time scale in the political system that is extraordinarily short, when one considers that decisions in this area should be not only for the lifetime of our children, but also for their children's lifetime. This ought to be the scale of thinking, and it isn't at all. Once in our society it was.

I think this is a serious problem, and a very complex one that deserves a good deal of attention. I have done some writing about it, but it's not something for which one can toss off a quick solution. I think we ought to be able to make decisions relevant at least to the lifetime of our children—which is not really very far ahead.

The third problem is that we have gradually built up a system in which there are thousands of places in which you can veto doing something, but we have no real organized method of going about setting priorities, no system of tradeoffs. We can block things all over the place, and every year we invent new ways, either legislatively or with the massive numbers of lawyers we're turning out. At some point, society has to be able to say we want to do all these things but we can't, or this is more important than that, or we can go so far on this and so far on that. We have very crude ways of accomplishing this, but we really need a much better approach to setting priorities and establishing tradeoff mechanisms. We really don't have a workable system, and that is a very serious shortcoming in our society. But how do you accomplish these things and stay away from being authoritarian? It's very easy to find an authoritarian solution, but it's very tough to find a democratic solution. That's what we need.

Another facet of this issue is the problem of a coordinating mechanism. The United States was built on the premise of free enterprise. We don't have a completely free-enterprise economy, but we are moving in that general direction as a country even though government has been getting more and more involved over the years. The problem is, how do we get to the point of trying to bring about some tradeoffs and coordinating when arriving at public decisions? Take, for example, in the field currently being described as industrial strategy, the single problem of how we as a society can stay ahead in the field of computers and automation. We are ahead in it, but a lot of very complex issues affect what happens next, and we don't have a very good coordinating mechanism for handling the implications. The communications debate at the moment is a very good example. We don't have a good mechanism for arriving at public policy in this area and for taking account of the many variables that will affect our society's future and its well-being. How do you do that and still keep flexibility and freedom?

We don't know. It certainly ought to be possible to do so. I think that in many ways it's easier to do than to change short-term time horizons to longer-term ones. But we do need to find a way to achieve this. Other societies do it in their way. Both the Germans and the Japanese are able to do a better job in this respect than we do. The French go at it in their way. I don't think we should copy any of them, but I think devising our own way is a worthwhile problem on which to be spending time. It certainly affects our ability to look ahead for more than a very short period of time.

Another area where we ought to be innovating is in really trying to understand the impact of discernible future change on current planning decisions. Some years ago, I suggested that we create some autonomous institutes of the future, which would be publicly funded but not tied to current budgets, and which would be insulated with public boards not related to political administrations in power. Having a number of such institutions would make it possible for some of them to take contradictory positions on particular issues relating to the current impact of discernible future change, and for the future impact of current decisions. I still think this would be a very useful thing for our society to do.

The last problem I want to point out here is that we need a guiding vision. The sense of aimlessness many people feel today hinders our ability to restructure our institutional relationships so that we can marshal our strengths. As a contrasting example, in the late eighteenth and early nineteenth centuries we had a sense of where we were heading as a young nation, an inspiring set of ideals that helped give us a common direction, and a view of how to draw together all the various elements involved in our growth at that time. We have at least as much individual wisdom today as we did back then, considerably greater knowledge, and far more powerful means at our disposal—if we could only use them as well! I am not suggesting that we try to resurrect or imitate all the values of the past. Rather, we need to find a new guiding vision appropriate to today's concerns. We also need to do this without decreasing our ingenuity or our individual initiatives, and without getting ourselves into a structure that is undemocratic or that goes against our history.

CONCLUSION

All these issues pose problems. I believe that the institutional change, the institutional inventing, and the political questions are vitally important. They will have enormous impact on whether the current technological and social revolution will affect future society for better or worse. There is very little going on in these areas that is encouraging. There is a lot that isn't.

Some people are trying very hard to come to grips with these issues, but I think many more people should be concerned. We need to rethink (to use the term that I used in this book) the institutional relationships within our society, both because technology has outpaced them, and because outmoded structures in many areas are holding back human forces—which are so important in putting to good use the technology we've created.

Our most important task in applying technology and science is to make sure that we use them as imaginatively and constructively as possible, so that all levels of society will benefit from them.

Preface

John Diebold, 1952

IF THIS book succeeds in indicating some of the ways in which we can more effectively use the technology of automation, it will be as a result of the countless hours many people have spent in talking with the author, in corresponding with him, and in demonstrating to him their work, their problems, and their plants. If this book does not succeed, it is because the author has failed properly to draw together the rich fund of material that has been made available to him and to extract from it the meaningful conclusions.

In either case an enormous debt of gratitude is due to far more people than can possibly be thanked here. Yet, certain people more than others have been responsible for making this book possible. It is the author's duty and privilege to make clear his debt of gratitude.

Brigadier General Georges F. Doriot has had more to do with this book than he will admit—for he, as keenly as any man, realizes the limitations of books. Yet it was because of General Doriot's imaginative understanding and kind help that the author was able to convert a long-time interest into a concrete research project at the Harvard Business School. That project was the starting point of this book.

The author's associates in the Research Group on Automatic Control Mechanisms at the Harvard Business School were J. Carlin Englert, Harry M. Gage, Jeffrey L. Lazarus, Jr., Melvyn A. Saslow, Nicholas C. Siropolis, Irwin M. Yanowitz, and John Wright. Thanks are extended to these fellow graduate students who shared a common interest during the writing of *Making the Automatic Factory a Reality.* The work of Irwin Yanowitz, on automation in the steel industry, has provided the basis for the discussion of steel making that appears in Chapter 3.

Mr. Edwin O. Griffenhagen and Mr. James G. Robinson have made the writing of this book possible as an unbroken project. As the general partners of Griffenhagen & Associates they have throughout extended aid and made available the full resources of the firm. All members of Griffenhagen & Associates have been kind and helpful, but especially warm thanks are due to my good friends Edward J. Harrington and Hugh J. Reber.

Marian Lee Ray has from the start given every imaginable help in preparation of drafts and in final typing of the manuscript. She has been assisted in this by Patricia Anand, Dorothy Bockman, Patricia Crawford, Marion DeMaio, Elizabeth Kulisek, Katherine Zurian, and Jeanne Crane Stuart.

If it had not been for Frederick G. Melcher—the *spirit* as well as the editor of *Publishers Weekly*—this project would never have materialized in its present form.

With an increasing sense of inadequacy, I have been perplexed by the problem of how even to begin to give full credit to the contribution made to this book by my wife, Doris. The answer, which of course should have been clear from the start, is that one can never adequately express all that is meant by the phrase, *a good wife.*

The Word *Automation*

John Diebold, 1952

AUTOMATION IS a new word denoting both automatic operation and the process of making things automatic. In the latter sense it includes several areas of industrial activity such as product and process re-design, the theory of communication and control, and the design of machinery. The connotation intended is delineation of these otherwise loosely related studies as being a distinct area of industrial endeavor, the systematic analysis and study of which will yield fruitful results.

The origin of the word is humble indeed. During the writing of the Harvard report, *Making the Automatic Factory a Reality*, the author found *automatization* both awkward and—from the standpoint of his weak spelling—hazardous. To be sure, there was also a growing appreciation of the advantage of recognizing the area of automation as distinct from the technology of control. But it is only fair to confess that it was the ease of spelling that finally overcame the author's reticence to coin a new word.

It has recently been brought to the author's attention

that Mr. D. S. Harder, Vice President in charge of manufacturing of the Ford Motor Company, has for some time used the word *automation* to describe the automatic handling of materials and parts in and out of machines. It is a pleasure to recognize here the efforts of a kindred spirit.

1

The Problem of Automation

IN 1784, before the industrial revolution had really begun, Oliver Evans built an entirely automatic factory just outside Philadelphia—a continuous process flour mill. Evans' mill made use of all three basic types of powered conveyors in a continuous production line. No human labor was required from the time the grain was received at the mill until it had been processed into finished flour. And in Paris, in 1801, Joseph Marie Jacquard exhibited an automatic loom controlled by punched paper cards, similar in many ways to the punched cards used in modern office equipment. Jacquard's loom became so popular that by 1812 there were eleven thousand operating in France alone.

Automatic controls in some form have thus been in existence since the beginning of the steam age. In fact, the governor on the steam engine was itself one of the earliest automatic control devices. As the steam age progressed and the industrial revolution got into full swing, the number of automatic machines increased. By 1833 even biscuit making at the "victualling office" of the British Navy had been mechanized.

Until recently the development of automatic control devices has been nonetheless sporadic, and automatic mechanisms have been built for the performance of only a limited number of tasks. During the last decade, however, developments in the fields of electronics, communications, and electric network analysis have made possible the construction of a wide variety of self-correcting and self-programming machines. These machines are capable of automatically performing a sequence of logical operations, similar in many ways to the mental processes of humans; they can correct the errors which occur in the course of their own operations and can choose, according to built-in criteria, among several predetermined plans of action. These recent advancements have been of such importance that they constitute the first stages of what coming generations will look upon as a second industrial revolution.

OVEREMPHASIS ON CONTROL

Paradoxically, the current obsession with the novelty and spectacular performance of automatic controls diverts attention from the problems of their application to industry. Although automatic control mechanisms are *necessary* for the achievement of fully automatic factories, they are not *sufficient* in themselves. The full promise of the new technology cannot be realized so long as we think solely in terms of control. Product and process redesign; the analysis of processes in terms of functions rather than in terms of the steps that are now being performed; rethinking of the entire operation—all these are necessary, important, and often infinitely difficult problems which must be solved before business can begin in any real way to take advantage of the Aladdin's lamp which technology holds forth.

Yet the amount of attention that has been given to the control problem has kept nearly all discussion of automatic industrial processes revolving about the control devices and control systems. The few times that the other problems have even been mentioned they have been voiced in an "aside," while the subject of the main address has been "control."

It is not difficult to understand why primary attention has been focused on the control aspect of the automation problem. It was the electronic and communications research of World War II that lifted the idea of a fully automatic factory from the realm of science fiction into that of serious discussion. The development of high-speed digital computers, or "giant brains," as well as mechanisms for directing guns upon rapidly moving aircraft, provided solutions to some of the most important basic automatic control problems.

It was apparent to those who planned these military machines that important changes would be possible in industry through application of the new technological principles. For the most part, however, the scientists working on the military devices realized that much work was needed in areas other than control in order to use effectively the new developments in automatically controlling industrial machinery and processes, but their major concern was with the control problems of the military and they rightly concentrated on those problems.

LOOSE THINKING

The popular imagination has been captured by the automatic control of guided missiles and by the electronic computers. These uses of control devices incite journalistic fantasy, for it is all too easy to draw

3

superficial parallels between the operation of certain military equipment and the operation of industrial machinery. Most of what has been published about the peacetime industrial use of the new technology, even when not outright science fiction, has presented the problem as primarily one of control. The result has been diffuse and loose thinking.

The confusion has not been lessened by the analogy that has been drawn between the operation of certain control systems and the operation of human and animal nervous systems. While this comparison may be useful for pedantic purposes, it applies at best to only limited areas. There are close analogies between electronic circuits and the nervous system, but the resemblance is frequently overdrawn. Widely publicized statements by responsible people that computing machines have *nervous breakdowns* and respond to *shock therapy* add to misunderstanding.

The result of these circumstances is that the businessman is kept thinking in terms of the device through the development of which the technological advances have been made. He thinks of the large high-speed digital computers and of the gun-directing mechanisms. Very little is said in business literature about the underlying techniques that are the really important part of these new devices, and that can permit the design of entirely new industrial equipment just as they have led to the development of entirely new military equipment. What the businessman needs to know is how the new developments can be applied to industry.

High-speed digital computers, although far less important industrially than many other forms of the new technology, have received most attention. But when a businessman is told that an immense digital computer

4

can multiply forty-two nineteen-digit numbers per second, and costs three and a half million dollars to build, he is clearly not being presented with the kind of information he needs. In addition, he is probably frightened away from any further investigation of the field of automation, despite the glowing pictures painted by those who say that smaller computers will soon be available much more cheaply.

The businessman is quite aware that, however small and cheap the new mechanisms may become, there is much that must be changed in *his* plant before electronic circuits are going to make automatic production possible. But, as yet, only scant attention has been paid to the problems he faces. Many times these problems have been attacked piecemeal, often without being recognized as essential parts of a distinct process—automation. Only in isolated cases, such as at the Ford Motor Company, have these problems been approached at all comprehensively.

It is the purpose of this book to describe these problems, and to show what benefits can be gained from studying them systematically. Once this has been done a basis will exist for realistically examining the economic and social effects of these important technological developments.

ECONOMIC AND SOCIAL EFFECTS

Some of the science fiction accounts of automatized industry have presented a grim picture of the future, with jobless and debased workers roaming the streets of a fully mechanized civilization—even receiving their relief checks from machines. All too many predictions of more serious-minded people have been equally black. The reason for this state of affairs is not hard to understand. As has already been indicated, the full extent of the problem of using the new technology in industry has not been made clear by the literature describing the automatic control devices. Few distinctions have been drawn between the problems of designing electronic control circuits and the problems of using these circuits in the plant. As a result the application of automatic controls to all forms of industry has seemed imminent. This supposed "fact" readily trips the trigger for a (perpetually loaded) double-barreled charge of social criticism.

However, very few analyses of the nature and extent of industrial automation have been made in terms of the realities of either our present technological knowledge and economic environment or of historical perspective. Yet it is meaningless to discuss the economic and social effects of automation without first determining the direction and extent to which automation itself will progress.

Automation will appear as a distinct phase in industrial progress, but it is nevertheless a part of the long continuum of man's mechanization of his work. The economic and social effects of the new technology should be viewed in this perspective. Although the present book does not present an exhaustive or systematic analysis of the social and economic effects of automation, an attempt has been made to discuss at some length the most important, and thus far the most misconstrued, of the many important questions in this sphere.

SCOPE OF THIS BOOK

Before the full benefits of the new technology can be realized, difficult engineering and business problems must be solved. This book is primarily an essay on

6

the *business* problems of automation and indicates the manner in which technological developments can be useful to the businessman. It points out the obstacles that will confront the businessman in making use of automation and suggests courses of action by which these obstacles may be overcome. Finally it discusses the important economic and social effects of the new technology.

In no sense does this book pretend to be a contribution to the body of technological knowledge on the subject of control. But this should not be interpreted as minimizing the importance of the technological problems that have been solved and that remain to be solved.

2

Control and the Computer

WE SHALL begin at the heart of the matter—
the theories of communication and control. But we shall
explore these subjects in purely functional and not technical terms.

COMMUNICATION THEORY AND CONTROL

During the last decade considerable effort has
been devoted to the formulation of a fundamental theory
of communication and control. Based on the relatively
new concept of the *bit* (or elementary discrete unit)
of information, this theory has been mathematically
formulated by Claude E. Shannon and W. Weaver in
The Mathematical Theory of Communication (published by the University of Illinois Press in 1949). The
considerable body of work done during and since World
War II on servomechanisms and control systems (treated
more fully later in this chapter) in conjunction with
this theoretical foundation of communication theory
permits the design of workable automatic control systems for performing a very wide variety of functions.
We are thus no longer dependent upon the sporadic

ingenuity of individual inventors for the automatic control of our industrial processes.

The basic contributions of communication theory in the last decade have been the concept of the *bit*, or the basic indissoluble unit of information, and an analysis of the transmission of these units of information in the presence of *noise*. Since noise, or disturbances in the meaningful patterns of bits of information during their transmission, may be caused by human phenomena as well as by electromagnetic static, communication theory is applicable to social organizational problems. (If the reader is not familiar with the central part played by the transmission of information in the effectiveness of social organizations, he would do well to read Chester Barnard's *The Functions of the Executive*, published by the Harvard University Press in 1948.)

Simultaneous and related to the work in communication theory has been the development in control theory of techniques for designing stable *feed-back* (or self-correcting) control systems for a wide range of uses. Such *self-correcting* controls had been available in the past, but only for a limited number of tasks. The failure to operate in a stable manner had been a chief deterrent to their widespread use. As a result, however, of extensive scientific analysis both in this country (especially at the Massachusetts Institute of Technology) and in Britain, these *closed-loop, self-correcting,* or *servomechanism* systems can now be designed to conform to rigid performance specifications and can be constructed to operate in a stable manner.

The work in both these fields, communication and control, together with the electric network analysis necessary to the development of radar, has made possible the design and construction of fully electronic digital computers supplanting the mechanical computation mechanisms and the differential analyzers first conceived by Charles Babbage in the nineteenth century and perfected by Vannevar Bush and his associates prior to World War II. It is these electronic digital computers which will eventually supplant many of the functions now performed by human beings in industrial control. But it is the perfection of *stable* closed-loop control systems which plays the most important, as well as the earliest, role in automation. And it is to a short, functional description of *closed-loop* control that we turn before describing the computers and to the important differences between closed-loop and open-loop control.

OPEN- AND CLOSED-LOOP CONTROL

Luckily, it is not necessary to draw a rigid distinction between the control mechanism and the machine being controlled. Such a process can be very easy, but it can also be uselessly difficult. For the purposes of this chapter an explanation of control in simple, functional terms is all that is required.

Control can be as simple as turning a machine or a device on and off, as, for example, an electric light or a kitchen ventilating fan. *On* and *off* may be the only control which we exercise. If that is the case, the device either works or it does not. With higher forms of automatic control, however, the fact that the machine does not work, or works unsatisfactorily, can be made to have an adjusting effect on the energy input to the machine or on the subsequent action of the control mechanism itself. In the simplest type of control, however, such adjustment must be made by the people using the equipment. In the kitchen fan, for instance, if it is an extremely hot day and a great deal of cooking is being done, the fan still turns at the same speed and moves the same quantity of air as it does on a relatively cool day. Here, the operation of the control, that is, turning the fan on or off, is independent of the action (cooling) of the system being controlled, except through the control exercised by the person in turning the fan on or off. The important distinction is that the control mechanism itself is independent of the performance of the system which it controls. This is called an *open-loop* control system and is perhaps best explained by means of a simple schematic diagram.

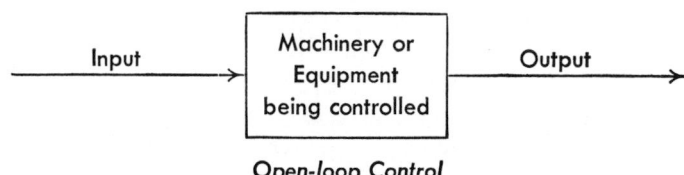

Open-loop Control

The open-loop control system is the simplest type of control process, and it has all of the basic elements of man's control of machines. Many of what we normally class *automatic* controls are open-loop systems. For example, an outdoor lamp which is turned on by a preset clockwork mechanism at the same hour every day (or at a different hour every day, depending upon the time of sunset), although automatic, operates independently of the condition of illumination. If an afternoon thunderstorm darkens the sky, the light does not go on.

IMPORTANT CHARACTERISTICS OF CONTROL SYSTEMS

Many open-loop systems, however, do possess one of the most important characteristics, from the standpoint of automation, of modern control systems—control at low energy levels. The energy expended in turning a switch or in opening a valve need have no particular relation to the energy that is being controlled. The flick of a switch can detonate a charge of dynamite. It can control the output of a large power plant. Or it may merely turn on a kitchen fan. It is not necessary to build complicated automatic control devices in a form capable of carrying high flows of energy. This fact is of vital importance. If this were not true, it would be entirely impossible to build many of the most advanced control systems.

Another important characteristic present in simple open-loop control systems is that the control element can be placed at a great physical distance, if need be, from the element being controlled. Detonating a TNT charge by the pressure of a key in the President's office for the opening of work on a dam two thousand miles away is a spectacular example. More useful, however, is the fact that this characteristic makes possible the grouping of all the controls of a chemical process plant in one location so that a fully automatic control system can be built.

Thus, open-loop control systems can possess two of the most important characteristics of the modern automatic control systems, control at low energy levels and control from a distance. But the most important control characteristic from the standpoint of automation, the ability to correct errors, is not present in open-loop systems. Indeed, control in terms of the operation of the system rather than entirely in terms of predetermined arrangements is the distinguishing characteristic of the other major family of control systems—*closed-loop* control. It is with the industrial phenomena made possible by use of closed-loop control systems that this volume is primarily concerned, but description of an open-loop system was necessary in order to make clear the distinction between the two types of systems. For automation is possible only through use of the recently acquired ability to design and construct a wide variety of closed-loop control systems.

The thermostat in your living room is, like the switch on the kitchen fan, a control device. Yet the living room thermostat is representative of a very different type of control system, for it is part of a closed-loop control system. In a closed-loop control system, the control device does not operate independently of the performance of the system or machine being controlled, as it does in an open-loop system; rather, the operation of the control device is, in part, a function of the actual performance of the system or machine being controlled.

For example, in a properly operating home heating system the human control that is exercised on the thermostat is the setting of the dial at the desired room temperature. If the temperature of the room drops below the desired point, the heating equipment is automatically started by the thermostat (or fuel is fed more rapidly into the furnace, depending on the system and operating conditions). When the room temperature rises to the desired point, the thermostat automatically turns off the furnace. (Actually, the thermostat operates over a range of several degrees of temperature. It does not turn the furnace on unless the room temperature drops a few degrees below the desired value, and does not turn it off again until it has raised the temperature a few degrees above the desired value. This is done in order to eliminate constant starting and stopping of the furnace. It is common to much of the automatic control equipment that ideal action is not practical. But this qualification is, in itself, not important to an understanding of the idea of self-correcting controls.)

An electric oven, depending upon its construction, may be an example of either an open- or closed-loop control system. If the oven is not equipped with thermostatic control, the heat which it gives depends upon the fluctuations in the electric current, the amount of prior operation, and even whether the door is open or closed. The temperature prevailing within the oven at any particular setting of the control dial has no effect upon the flow of current. This is open-loop control. If, however, the oven is equipped with a thermostat (as are most recent models of electric ovens) , the temperature in the oven is held to within a small predetermined range of temperature. If the oven has operated for several hours previous to cooking dinner, less current is needed to heat it to the desired temperature, and the control mechanism cuts down on the amount of current flowing into the heating coils. Likewise, operation over a long period of time, with a thermostatically controlled oven, will not result in raising the temperature above the desired range.

The human use of tools and machines (which in themselves have only simple open-loop controls) depends upon a closed-loop system of control. It is the human muscular and nervous system that provides closing of the control loop. The process of steering one's automobile is an everyday example of this sort of control. The steering mechanism is not an *automatic* closed-loop control system, but the process of steering an auto contains all of the important elements, and some of the special characteristics, of a closed-loop control system. The driver provides the closed portion of the loop by controlling the equipment in accordance to operational requirements.

When you are steering your automobile, the direction of the automobile is controlled by your handling of the wheel. When you arrive at a turn in the road, it is by means of the steering wheel that you direct the automobile to follow the turn of the road. If, however, the wheel is turned too far in the direction in which the road turns, the car will tend to go off the side of the road; or if you do not turn it far enough, the car will go over the center dividing line. Actually, of course, what occurs is that you start to turn the wheel and realize it is not quite far enough and increase the turn. Or you realize that you have turned it too far and decrease the extent of the turn. Your eyes perceive the position of the car in relation to the road, and you realize the amount of error, as well as the rate of change of error. A correction signal is sent from your brain to your arms and you make the necessary adjustment. An experienced driver accomplishes these necessary corrections without conscious thought. Such action is an ideal example of a closed-loop control system. The original energy input from the control device, that is, the driver, is a predetermined notion of the amount of turn needed in order to keep the car in its correct position in relation to the road. If the degree of turn of the road was not perceived correctly, or if it is a wet day and the wheel turns more easily than was expected, the amount of energy needed is different from that which was put into the control mechanism (the steering wheel), and the car turns either too far or not far enough. But a correction signal, initiated by the driver's eyes, provides for alteration and correction of the original energy output. Thus, the operation of the control mechanism is intimately connected with the operation of the device being controlled.

It is interesting here to note another similarity between the driving of an automobile and the closed-loop control of certain machines. The phenomenon known as *hunting* in closed-loop systems occurs when too great a degree of correcting energy is put into the control device. As a result, the mechanism overshoots the desired point. In the example of steering an automobile, an inexperienced driver may turn the wheel too far in the direction of road turn. The car starts to move off the road. In hastily attempting to correct his action, the driver turns the wheel too far in the other direction and the car moves too far to the other side of the desired direction. Similarly, in some machines, if the mechanism overshoots, the entire machine can be thrown into a state of unbalance, for the correction mechanism will attempt to bring the object or quantity being controlled back to the desired position, and very often again overshoots the desired point. Changing requirements of energy under different operating conditions could cause such "overshooting" or "undershooting." For example, when driving on a wet road, less energy may be needed to turn the wheel. Expenditure of the normal amount of turning energy, under these conditions, results in the wheels' being turned too far.

It is the developments of the last decade in control theory, permitting the design of a proper amount of *damping* (or resistance to immediate correction) into the system, which have made it possible to design stable closed-loop control systems. Formerly (even now in an improperly designed system) the process of *hunting* frequently resulted in progressively larger deviations, the violent action of which often caused the machine to fail.

The following diagram illustrates the manner in which a closed-loop control system is constantly corrected by the quantity of error between the actual state of operation and the desired state of operation.

Closed-loop control is not new. What is new, and what the developments of the last decade have made possible, is the application of closed-loop principles to the control of a great variety of man's endeavors. It is

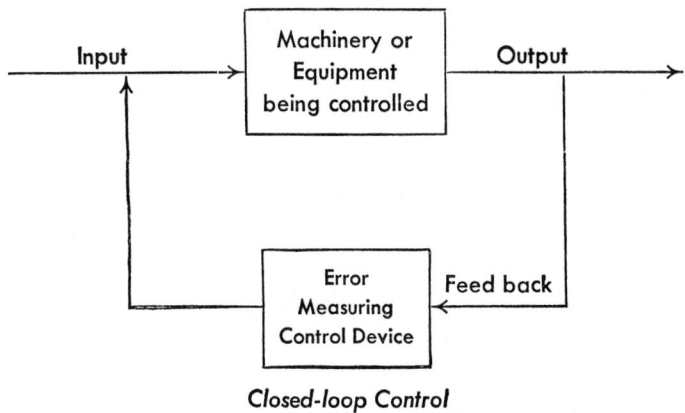

Closed-loop Control

worth while, however, to sketch briefly the early history of closed-loop control systems, for with proper perspective the nature and importance of the most recent developments become clear.

THE DEVELOPMENT OF CLOSED-LOOP CONTROL

The regulation of the speed of machines was the earliest application of closed-loop control systems. And it is in terms of speed regulation of machines that the earliest theoretical analyses of closed-loop control systems were made. At the end of the eighteenth century James Watt developed a centrifugal governor for steam engines for the purpose of maintaining a constant engine speed, regardless of the amount of work that the engine did. When the engine tended to slow down, the centrifugal governor, which consisted of two rotating metal balls at the end of a simple lever arrangement, allowed a greater flow of steam into the engine. When the engine was required to do less work, less steam was necessary to keep it turning at the set speed and the governor partially closed the valve that controlled the flow of steam to the engine. Most of the work on control systems during the nineteenth century was in the form of such speed regulators. Toward the end of the century, the closed-loop control systems began to be analyzed theoretically. James Clerk Maxwell, and the astronomer Sir George Biddell Airy, wrote on the theory of closed-loop controls during this period.

Position control (also known as *remote position control* or *r.p.c.*) made its first appearance in the form of control systems for the steering gear of ships. The amount of energy necessary to move the rudder of a ship to a desired position before position control was effected varied with the condition of the sea and the speed of the ship. Yet today under modern steering engine arrangements a turn of the ship's wheel by the pilot requires an unvarying amount of energy. The actual movement of the rudder is accomplished by an engine in the stern compartment of the ship. The movement of the wheel by the pilot acts as a control for this engine. Variations in the amount of energy (steam) necessary to turn the rudder are provided by means of a mechanical linkage system. This system, developed late in the nineteenth century, represents the first form of closed-loop *position control*. Here, the steam flowing into the engine (which moves the rudder) is controlled in terms of the actual position of the rudder. If it is necessary to move the rudder farther, more steam is admitted. If the rudder has moved too far, the action of the rudder is reversed.

In 1922 Minorsky published a theoretical analysis of the position control system of a ship's steering engine. But modern closed-loop control work did not really get under way until the decade of the 1930's. During that period attention moved from regulators to control of processes such as the refining of oil and the manufacture of certain chemicals. The first thorough treatment of servomechanism theory was given in 1934 by Hazen, earlier work having been done by Wyquist in 1932, who first showed how stability could be determined in servo systems.

However, it was the military demands of World War II which gave the greatest impetus to the development of servo systems (or closed-loop systems). The speed and maneuverability of aircraft quickly made traditional gun-laying techniques obsolete. In order to meet the pressing military requirements, considerable numbers of highly capable scientists went to work on the problem of designing and constructing automatic position control, or servomechanism, systems. Out of this work came a substantial body of new technical knowledge. It is not necessary here to describe this complex work. Its meaning for industry is that stable servomechanism, or closed-loop, control systems can now be designed and constructed for performing a great variety of tasks. What was once available only on a limited basis— control of a ship's steering gear or of the speed of a machine—has now been analyzed and developed to such a stage that the principles can be employed in a scientific manner for the solution of a wide range of control problems.

In addition to the development of control systems, the recent attention to servomechanisms has contributed a very useful by-product to industry in the form of the techniques that were developed for solving servomechanism design problems. These *servo techniques,* as they are called, provide a useful tool of analysis to engineers studying the characteristics of certain industrial processes. By building *analogues,* or models having characteristics similar to the process to be studied (although quite different in form, i.e., the *model* may be an electrical network), it is possible to create artificial disturbances in the operation of the model and study the characteristics of the actual system's reaction to these conditions. It is in this way possible to *experiment* with an industrial process economically and quickly and without running the risk of damaging actual equipment or of slowing production.

The Brown Instrument Division of the Minneapolis-Honeywell Regulator Company has done considerable work on the use of servo techniques for the solution of problems in the design of industrial process systems to take maximum advantage of automatic control mechanisms. (Technical papers, both by Minneapolis-Honeywell engineers and by others working on these problems, are available to those wishing to pursue the analytical use of servo techniques.) Perhaps the most interesting ramification of this work is the ability to design industrial processes around automatic controls. Considerable work has recently been done in the chemical process industries to design processes "around the instruments."

THE COMPUTERS

Partially as a result of military needs for trajectory and ballistic tables and for information of a similar nature, and partially in fulfillment of a long-sought-after scientific goal, World War II also produced a new type of electronic machine, the high-speed digital computer.

It is around the computers, or "giant brains" as they were quickly dubbed, that most of the literature and the fascination of the new technology centers. The importance of these new machines to future industrial automation is indeed great. Unfortunately, however, this importance has been somewhat overemphasized; or perhaps it is only that the industrial problems which remain to be solved have been minimized.

Whatever the reason, something of the opinion seems to prevail—at least in magazine articles, and little else is available to the layman on this subject—that it is the computers which provide the key to the fully automatic factories of the future, and that practically all that is necessary to make factories automatic is to find a cheap way of building computers. Although it is obvious to anyone who works in a factory, or who seriously thinks about the problems of fully automatic factories, that automatic materials-handling problems, for example, present far more formidable stumbling blocks to automation than do automatic control systems, little formal recognition has been given to these other problems, and the computers go on being hailed as the answer to automatic production with the other problems treated as peripheral in nature.

Despite this almost complete neglect of other pertinent and important problems in favor of the fascination of the computers, it is not intended here to tip the scale in the other direction and detract from the importance of the development and perfection of these machines. Combining many of the most important technological developments of the last decade, the computers embody the new technology of control in its highest form. They represent a family of machines already serving mankind. With proper development (and the solution of the other problems of automation), they will, in the future, have much effect upon our way of life.

Although the word *computer* is applied to many devices, including desk calculators, it is primarily coming to mean machines that, once set up, perform a *series* of individual computations or steps in the solution of a problem, without further human intervention.

The idea behind these machines was well over 100 years old when the first of the new all-electronic digital computers was completed in 1946. In an effort to overcome the errors in astronomical tables, the English mathematician Charles Babbage outlined in 1822 the first *difference engine,* a mechanism comparable to today's computers. The machine, which Babbage received government funds to build, was remarkably like our present-day computers conceptually, although quite different physically, for it was mechanical rather than electronic. The technical competence of his age was not sufficient, however, to permit completion of a working model, and after a number of years Babbage turned his attention to design of a more advanced *analytical engine.* This effort too was thwarted by the limited machining methods of the early nineteenth century.

Later in the same century William Thomson (Lord Kelvin) suggested solving mathematical problems by means of machines and described the form which integrating and differentiating components might take. Again, tools capable of the required precision were not available; the mechanisms were never built.

In 1925 Vannevar Bush and his associates at the Massachusetts Institute of Technology turned to the problem of designing and constructing a machine capable of solving differential equations. They worked with modern machine tools, and by 1930 they had succeeded in building the first workable *differential analyzer* which solved problems entirely mechanically, using electric motors to drive the mechanical parts.

During the 1930's more machines were built for the mechanical solution of mathematical problems. Although such devices represented great strides in the handling of large mathematical problems, they had definite limitations, chiefly that considerable mechanical changes had to be made in order to adjust the machine for the solution of different types of problems. In addition, although they permitted far more rapid solutions than had been possible previously, the speed limitations of any machine depending upon levers and wheels still left much to be desired.

The electronic developments of the 1940's made it possible to perform by means of electronics what the differential analyzers did mechanically. With the completion of the IBM *Automatic Sequence-Controlled Calculator* (Mark I) at the Harvard Computation Laboratory in September, 1945 and the *Electronic Numerical Integrator and Calculator* (ENIAC) at the Moore School of Electrical Engineering, Pennsylvania University, in February, 1946, a new family of modern high-speed digital computers came into being. There are now about forty of these machines in operation or under construction, and such characteristics as speed and capacity have been improved enormously by the considerable scientific and engineering talent that has been, and is being, expended in what is virtually a computer building boom.

Although the original reason for building computers was to provide a means for solving mathematical problems, it became evident with the construction of the electronic machines and the advances in communications theory that information other than mathematical data could be processed by the computers. The enormous speed of these machines and their ability to follow extensive lines of logical reasoning and to choose between alternatives (upon the basis of built-in criteria) make them ideally suited to the handling of large quantities of routine, and not so routine, business information. With this realization, the idea of fully automatic factories and offices entered the realm of industrial possibility.

Much work remains to be done to adapt computers to the requirements of business. Most existing computers have been designed to solve a high order of mathematical problems occurring more frequently in science than in industry. The advances made in designing the so-called *scientific* computers, however, have helped greatly to simplify and improve circuits adaptable to smaller machines for business use. Companies have been organized expressly for the purpose of building such *industrial* or *business* computers. It is important for the businessman to know what computers do, for until he does, descriptions of their size, speed and cost have little meaning.

There are two classes of computing mechanisms— *analogue* computers and *digital* computers. Analogue machines solve problems by creating physical analogies to the problem being solved. Quantities are represented by means of physical distances, angles of turns, or varying electrical voltages. An analogue machine measures and uses these quantities in much the same way an individual using a slide rule solves a problem by the relative position of the two portions of the rule. In contrast with this, a digital computer *counts* discrete units, or digits, rather than *measures* quantities. The accuracy of the digital machine is as great as the capacity of the machine for holding digits, whereas the accuracy of the analogue machine depends upon the precision with which the various measurements are made by the component elements.

THE ANALOGUE COMPUTER

The Bush *differential analyzer* is a mechanical analogue machine, as would have been Charles Babbage's *difference engine* had he possessed the tools with which to build it. For when it was necessary to design these machines entirely in terms of mechanical devices, such as wheels, screws, cams, and couplings, great skill and precision machining were necessary. However, it is now possible to build entirely electrical analogue computers which measure changes in voltage rather than changes in the position of wheels. Such machines are cheaper to construct and do not rely upon fine precision workmanship as did the mechanical analogues.

Analogue computers, especially the electrical ones, are very useful for the construction of models, or analogues, of industrial processes and systems, and for determining the reaction of the system to various operating conditions to which it may not be possible to subject the actual system or process itself. For example, an analogue of a water pipe network can be built by plugging together the correct electrical units, and the effect of varying operating conditions and loads studied easily, quickly, and with little cost or danger. The analogue computer, in various modifications, is thus the principal means for utilizing the servo techniques, the use of which in studying and designing industrial processes was mentioned earlier in this chapter.

The *digital computers* have figured, however,
most prominently in the new technological developments, and it is this form of computer that will be
of most use in controlling and, in the case of information processing, of actually performing the tasks of
industry.

Digital computers rely upon the basic arithmetic of
addition and subtraction (from which multiplication
and division are derived) for their operation. It is their
ability to perform basic arithmetic at fantastically high
speeds (in millionths of a second), and to control the
sequence of their operations automatically, according to
built-in and changeable instructions (called *taping,*
after the paper and magnetic tape which are sometimes
used), that make them of such great use both in the
solving of problems and in the handling of other types
of information. By performing a long series of what
might be called "elementary school" steps in very rapid
sequence, it is possible for a computer to solve highly
intricate problems in very short periods of time. Since
mathematical notations and numbers themselves are
simply symbols, and since the symbols we use in ordi-
nary communications, that is, words, can be altered to
correspond with the symbols which a computer is built
to handle, it is possible for computers to handle all types
of information. Although we talk in terms of a computer adding and subtracting, they can also compare,
collate, and make logical choices between alternatives.
All such choices, it should be made clear, are based
upon built-in and alterable instructions, the computer
having no semblance of free will.

It has been expedient to build most computers to

handle what are called *binary numbers,* rather than numbers based upon our common system of tens. Binary numbers are composed of only two digits, 1 and 0, and represent all higher numbers by combinations of these two digits. Thus, the computer handles only two symbols, 1 and 0, rather than ten, as in the decimal system.

Since similar combinations of 1 and 0 can be used to represent letters of the alphabet, it is basically no more difficult for computer circuits to handle logical problems represented by word symbols than to handle mathematical problems.

The physical construction of the computers varies greatly. Thus far virtually every machine is unique. For example, the *memory unit,* or method of storing information within the machine, may take the form of a rotating magnetic drum, magnetic tape, paper tape, punched paper cards, or cathode ray tubes. Yet *functionally* all computers contain very much the same kind of units. There is a *computing* element which actually compares data, or multiplies, or makes a logical choice. A *control* element schedules the sequence of operations and automatically programs the action of the computing unit. *Memory,* or storage, elements retain, in permanent or temporary form, the information necessary for the operation of the machine. In addition, it is necessary to introduce information into the computer and to withdraw information from the computer.

Since the primary functions of the modern high-speed computers are performed electrically, there are very few moving parts on the most recent design, except for the introduction and removal of information which requires the turning of rolls of wire tape or the sorting of punched paper cards. In fact, nearly all digital computers resemble nothing so much as a large, dial telephone installation (except that many are encased in streamlined cabinets). Since vacuum tubes, or valves, as the British call them, are relied upon very heavily in the electronic networks, the generation of heat during operation and the general bulk and weight of the equipment require that heavy metal frameworks be built to support the computer, and that the components be sufficiently far apart so that the heat of operation can be carried away by a large-capacity ventilating system. The computers thus typically occupy one or more large rooms. As *transistors* (tiny strips of germanium isotopes which perform the same function as our vacuum tubes) replace vacuum tubes in the construction of computers, the heat problem will become insignificant, since transistors emit a negligible amount of heat. The computers will also occupy far less space; a computer built entirely with transistors rather than vacuum tubes will occupy about the same cubic volume as a normal office desk.

In order to *program* a computer, or set up the operating instructions which the machine must follow, each problem must be reduced to a series of very simple steps. At each stage the machine must, figuratively, ask a question which can be answered by "yes" or "no" (represented by 1 or 0). If a large number of repetitive steps are involved in the solution of a problem, where great quantities of information must be processed in the same way, it is economical to perform the type of analysis necessary for programming. If, however, the problem is unique, and cannot be broken down into a series of repetitive steps, the cost of writing out instructions for the machine to follow will be greater than the cost of manual solution of the problem. Many of the decisions which we make unconsciously when solving even the simplest problem would fill pages when set down in the form of computer programming, with alternative courses of action listed at every step.

Once it is properly programmed, a high-speed digital computer can automatically process a great variety of information, and can take all necessary action in connection with the analysis of such information (if the proper equipment is connected to the computer itself). By means of automatic typewriters and accounting machines, a digital computer can be designed to perform all the accounting and bookkeeping functions of a business (as is explained more fully in Chapter 5). If it is attached to the controls of automatic processing equipment, a digital computer can monitor the operation of an entire manufacturing plant. By very rapidly "asking" a series of questions of each machine control system, and each automatic inspection device, the computer can quickly correct errors which have already occurred or which certain combinations of events indicate will soon occur. With such computer control it is possible to extend the industrial use of the principle of *end point control*—whereby a signaled change in the requirements of the end product automatically brings about the necessary changes at each point in the manufacturing process. Such control is now possible in a limited number of the already highly automatic process industries, such as oil refining. By use of computers it will be possible to introduce end point control in whole, or in part, to several other industries.

In summary, the digital computers provide a means of automatically governing the operation of much of our manufacturing equipment. Where processing equipment can economically be connected to permit a continuous flow of product units, entirely automatic plants are possible. In a far greater number of cases, *portions,* but not all, of the operations of a business can be put on an automatic basis.

When used with insight and ingenuity, computers will permit relief from the most repetitive form of human work. They will make possible more rapid and less wasteful methods of increasing our material well-being. By virtue of their great speed and ability to handle simultaneously many variables, they will permit us to perform many tasks which until now we could not reasonably attempt.

3

The Redesign of Product and Process

<hr>

INDUSTRIAL AUTOMATION brings new problems as
well as new solutions. To harvest all of its fruits we
must, in many cases, redesign our products, our processing methods, and our machines. And to redesign them
we must *rethink* them.

The most popular pattern of presentation of automatic control at the moment seems to be to divide the
problem into three areas: sensory mechanisms, which
act in a manner similar to the sensory organs of the
individual to detect physical occurrences; a central control area, similar to the human brain, which acts on
the stimuli received from the sense organs and determines what should be done; and the effector mechanisms, similar in many respects to the hands and fingers
of humans, which carry out the control orders sent by
the central control mechanism. Such analysis is all
very well if the problems raised by the construction of
suitable *sensory mechanisms* and *effectors* are made
clear.

However, the usual way in which this analysis has been used has been to say: ". . . the elements of the automatic factory are already with us; all that remains is to connect the proper instruments to the computer (the central control mechanism) and attach our machines." But this is simply not true. The current excessive emphasis on computers and control mechanisms is responsible for this inaccurate but prevailing idea. This unfortunate emphasis has limited both the direction and the nature of our thinking about automation. It obscures the fact that much important work remains to be done in other equally important areas. This work cannot be regarded as secondary to the design of control systems.

Redesign of product, or of process, or of machinery—and sometimes of all three—is often necessary in order to take full advantage of the new technology. It is both erroneous and self-limiting to think of the possibilities of automation merely in terms of connecting to a computer today's machines and making precisely the same products we make today in much the same way. There is an enormous difference between a process which merely *makes use of automatic controls* and a process which is *truly automatic.*

Such a distinction may be exemplified by the steel industry, the basic industry of our economy. Many automatic control devices are used in steel production but, with the single exception of continuous casting, steel production is not automatic. Nor is it likely to become automatic, regardless of how many electronic devices we employ, until there is a basic change in the way we make steel.

Photo tubes are used on Bessemer converters to determine the temperature of steel during the heating process. Similar devices are used for counting steel sheets as they are cut. Where wire or steel sheet is reduced in size by pulling it through dies or passing it between rollers, the successive motors must move faster as the wire or sheet becomes thinner; steel producers have solved this difficult control problem with automatic electronic voltage and speed regulators. Electronic devices are also used to determine when a particular heat of steel in an open hearth furnace is complete and when the furnace can be tapped.

Despite the fact that such a variety of automatic controls is used, steel is not made automatically. This is not due to reluctance to adopt automatic devices; it is a consequence of the process by which steel is made.

For years the steel industry has smelted ore and turned out ingots by using the blast furnace and the open hearth. Although electronic devices can be made to open and close the furnace doors and automatically to perform other operations at the furnaces, these devices scarcely tap the full potential of modern technology and in many cases do not even justify their cost. Fundamentally, use of the blast furnace and the open hearth is batch processing, just as oil refining was a batch process before the development of cracking towers. Over the years, continuous use of this method has built up such a substantial body of empirical operating data that steel men are hesitant to cast it aside for new and untried methods.

Two years ago I had the opportunity to visit one of our largest steel plants. At a long line of open hearth furnaces, the guide proudly explained the use of a new device for determining the optimum time for tapping the furnaces. As we watched, the workmen dutifully referred to this device and noted the readings on log sheets. Yet no furnace was tapped until an old foreman squinted through a piece of colored glass at a sample of molten metal that had been poured on the brick floor and nodded his approval.

It would be wrong to give the impression that changes in the technology of steel making are slow in coming about simply because of a dislike of new things on the part of steel makers. There are sound, and very obvious, economic reasons. The capital investment involved in the production of steel is enormous—as high as $35,000,000 for a single blast furnace with its auxiliary coke ovens and blowers. With such heavy commitments of capital and technical know-how, it is very difficult to make any radical departure from proven methods. Substantial changes in technology present too great a risk for most steel companies to accept during good times when they could afford it; for, as at present, they feel they are too pressed for production to experiment with radical changes. During depression periods they simply do not have the capital available for experimentation, and new security issues for such purposes are not well received.

In the words of Warren Kendall Lewis of M.I.T., "Our most basic industry has never submitted itself to a thoroughgoing scientific analysis; everything about it has remained empirical for a hundred years." This statement has been echoed by several of the most forward-looking steel men.

If it were possible to change the basic steel-making process as we have changed the process of converting molten metal into castings, we could perhaps contemplate automatizing the steel industry. Until then, the industry will simply be a production process that uses automatic control but is not fully automatic.

The single exception, continuous casting, has been automatized. It is perhaps worth while to look at this process more closely.

The casting of steel dates far back to early times when men made molds, poured molten metal into them and then removed the casting and processed it by hammering and reheating. In most casting mills the same procedure is followed today. Large molds are made; molten steel from either the open hearth or a Bessemer converter is poured into the molds; the molds are stripped away as soon as the metal is solidified; and the hot steel ingots are carried to the rolling mills where they are rolled into blooms and billets.

Steel men have long hoped and searched for a continuous casting process whereby, using a single mold, they could pour molten metal into one end and extract the hardened casting from the other. It was not until about ten years ago that continuous casting became a reality, and then only for nonferrous metals. At the outset the castings made by the continuous process were superior to the castings produced in the conventional manner.

The Scovill Manufacturing Company, a founder and prefabricator of brass, has been using the continuous casting process for some time with great success in the production of brass castings of uniform cross section. To the observer the process is deceptively simple. Molten metal is fed into one end of a long permanent mold. The bottom of the mold retains the molten metal until it starts to solidify. The mold is cooled by water to accelerate hardening. As the metal solidifies, the bottom moves out of the mold, and the hardened casting acts as a slowly moving plug to retain the freshly molten metal until it too has hardened. This results in the continuous formation of a single piece of brass, liquid at one end and hardened casting at the other, as it moves continuously through the mold.

Essentially the same process is used in the continuous casting of steel as is used in the casting of brass, but many problems were encountered in applying the process to steel making. Since the temperature at which steel melts is much higher than that at which brass melts, cooling the steel as it moves through the mold is more difficult. In addition, there is the problem of the reaction with air as the steel is poured into the continuous casting machine. Republic Steel Corporation and the Babcock & Wilcox Tube Company have been pioneers in the field of continuous casting of steel. In 1948 they announced a successful continuous casting process, using water to cool their molds, and a blanket of argon gas about the molten metal to prevent reaction with air as it is poured into the machine. Republic Steel and Babcock & Wilcox claim that steel produced by their continuous casting process is considerably better than the steel produced by conventional methods. The surface of the billets is freer from imperfections than that of ingots, and the interior of the billets is freer from slag. In addition, the continuous casting process eliminates handling the steel between the stages of formation of ingots and rolling of billets.

REDESIGN OF PRODUCT

It is often necessary to redesign the product as well as the process for fully automatic production. Sometimes the changes are small. For example, the addition of reference points on a casting may permit automatic positioning of the piece on a machine tool. Even with consumer goods, minor changes rarely affect consumer acceptance. The small glass nipple (or lug) on the side of a liquor bottle—which permits automatic positioning under the labeling machine—does not reduce product acceptance. As a matter of fact, redesign usually, but not always, is easier to achieve with consumer goods than with industrial parts that must fit other parts and perform precise functions. An ice cube designed with a hole in the middle allows fully automatic production and by providing a larger cooling surface is a better product. The consumer does not object to the hole. But how does one redesign a pretzel? It is much more efficient to stamp pretzels automatically out of a flat sheet of dough than to tie strips of dough into knots. But stamped pretzels do not have wide public acceptance. It was therefore necessary to design a machine that actually ties the dough into knots.* In this instance, special characteristics of the product, although they hampered automatic production, were a necessary element of the design and could not be changed. There are, however, many items like the ice cube, where product changes to achieve fully automatic production are perfectly feasible.

REDESIGN OF RADIO CIRCUITS

Printed radio circuits provide one of the best examples of a product redesigned for automatic production. Most radio circuits are composed of many small components wired together. Each joint where a contact

---

* The American Machine & Foundry Company designed and builds such an automatic pretzel tying machine.

is made requires soldering. Each wiring and soldering operation is done by hand. The complexity of the circuits is such that radio circuit assembly seemed to be a process in which automatic production was unsuitable.

If an attempt were made to automatize radio circuit assembly by mechanically reproducing the hand operations, a Rube Goldberg device of stupendous proportions would be necessary. This fact, coupled with the knowledge that the whole device must be flexible enough to permit continuous changes in the design of the circuits being produced, has been enough to make most people shy away from any attempt at automatizing circuit assembly. However, by thinking of circuits in terms of their functions rather than their present physical form, it has been possible to solve the problem by designing the wiring circuit in the form of flat planes.

The design of electric circuits in the form of flat sheets of conductive material attached to an insulating material is not new. In 1927 F. W. Seymour obtained a patent on a plated circuit. In 1929 H. H. Wermine patented a system for stamped wiring. H. G. Arlt received a patent on a sprayed circuit in 1937, and in 1939 a Swiss patent was granted to Heinsch for cast connections. Considerable attention has been given to the problem in recent years, because of the need for *miniature* circuits in the control of guided missiles. In fact, *printed circuits* (the generic term now taken as descriptive of most of this field) at present represent a rapidly growing, and most promising, development. Robert L. Swiggett, the Executive Vice President of Photocircuits Corporation, in an article in the August 1951 issue of *Modern Plastics* lists four principal methods in use at the present time for "printing" circuits. These are: (1) printing of silver ink on ceramics, followed by high temperature firing to fuse the silver; (2) spraying of metal into depressions in a plastic plate; (3) stamped metal patterns, either bonded to a plastic base or used as a rigid wiring harness; and (4) etching of metal foil-clad plastics.

A completed circuit may require several layers of printed elements. It is also necessary to attach components, such as vacuum tubes and resistors, at certain points. In most instances these attachments require hand positioning and connecting, although the soldering can be accomplished by any of several newly developed methods of dipping the entire unit into a bath of molten solder. Nevertheless, the design of circuits with printed, punched, or painted patterns is truly revolutionary and an ideal example of product redesign. Without such redesign, automation would be uneconomical, because it would require substantial investment in an elaborate assembly mechanism that would become obsolete with the first substantial change in circuit design.

AUTOMATIC PRODUCTION OF RADIO CIRCUITS

Shortly after World War II John A. Sargrove, a British radio engineer, perfected a machine for the automatic production of radio circuits by using a variation of the printed circuit for his basic design. Instead of designing circuits in terms of single planes, he used molded plastic plates that added a third dimension. In this way he provided the contours necessary for "building-in" certain of the circuit components that must be manually attached to the more elementary printed circuit. In Sargrove's circuits the plastic plate is more than the supporting device that it is in the simple printed circuit. Since the insulation plate is molded, it provides the contours that allow it to act as an integral part of the circuit. The plate can thus be used as the dielectric or insulating medium for electronic and electromagnetic couplings. Since the plates can be molded in any shape, a variety of circuits can be produced without basically altering the mechanism. This is one of the important advantages of the Sargrove process.

The automatic portion of Sargrove's process begins when the plastic plate is inserted at one end of a long multi-unit machine. In the first stage of the machine the plate is sprayed with an abrasive material in order to roughen the plastic surface. The plate is then inspected by photoelectric cells to determine if it is sufficiently rough. (If not, a signal flashes. If more than four pieces are rejected, the flow is stopped.)

The plate is then sprayed with molten zinc, after which it passes through a milling device where portions of the zinc are machined off. Throughout the system, checking devices stop the mechanism or discard the plate if mistakes have been made. In some instances, the mistakes are corrected automatically. Components, such as sockets for tubes, are automatically checked and inserted into the plate. At the end of the line, the finished plates are removed. There remains only a small amount of manual work in assembling a few plates, some radio tubes, and the amplifier to complete the radio receiver.

In the course of operation Sargrove's machine has had to be shut down, not because of any basic fault in design, but because variations in the properties of the component electric parts produced a high incidence of defects in assembled sets.* The defect is thus not inherent in the process and will be overcome as more rigid

* I am indebted to R. L. Meier of the University of Chicago for this information which is not generally known.

specifications become standard in radio parts. However, even if it were a fault which could not be corrected, the redesign of radio sets in terms of molded plastic plates sprayed with conductive material is certainly a notable achievement. But Sargrove did more than this.

In developing his machine, he solved some of the basic problems of the automatic factory. *His machine allows for product change.* A number of different circuits can be produced with only slight variations in the mechanism. Each processing stage—sandblasting, inspecting by photoelectric cells, spraying, milling, the insertion of sockets—is housed in an individual cabinet which can be separately removed from the line or altered as variations in the circuit design become necessary. With his machine Sargrove produced a radio which performs the same functions as a radio produced by conventional wiring. Admittedly, Sargrove's radio contained a very simple radio circuit, but his process is not limited to the production of simple circuits. Television and radar circuits can also be produced by similar machines. By what appears in retrospect to be a very simple redesign of product, Sargrove has made possible the automatic production of radio circuits—previously considered one of the most difficult processes to put on an automatic basis.

During the last few years a number of other radio and electronic circuit assembly problems have been solved by using variations of the printed circuit method. A recent paper by W. H. Hannahs and W. Serniuk * of the physics laboratory of Sylvania Electric Products gives an interesting example of the part product redesign can play in automatizing the assembly of such circuits. The authors describe a design for miniature

* *Electrical Manufacturing,* August 1951.

amplifiers in an airplane intercommunications system. Such units are necessarily very small, present difficult assembly problems, and require skilled production workers. In the new device, designed expressly for automatic assembly, all major components are reduced to a cylindrical form. Nine terminals are placed in line about the cylinder and connections are made automatically by fitting a wrap in place about the unit. An automatically indexing soldering machine permits one operator to assemble the entire circuit.

The amplifier performs the same functions as one assembled by hand but, through redesign for automatic manufacture, is much more economical than would have been the case if the assembly procedure had merely been mechanized to produce the original design.

OTHER EXAMPLES OF REDESIGN

Another interesting example of product redesign is found in the work of A. O. Smith Company engineers during World War II. The usual way of manufacturing airplane propeller blades at the beginning of the war was to contour-mill rough forgings of the entire blade. Contour-milling requires highly trained machinists and precise machine tools. The process was too slow for war production; many completed airplanes were grounded for lack of propellers. The engineers at the A. O. Smith Company, familiar with advanced metal-forming procedures, redesigned the propeller blades—not their shape, but their physical construction. By designing the blades in several parts that could be stamped individually on automatic presses and then welded together, it was possible to speed up the manufacture of airplane propellers, and to produce blades more cheaply than by contour-milling. Using a similar approach, engineers of the same company redesigned bombshells by using formed metal parts that were welded together to form the finished shell, eliminating slow and costly machining of castings.

Actually product redesign for ease of manufacture is not limited to the area of automatic controls or automatic production. It is a problem that is faced every day in a manufacturing concern. Often, it is not recognized as a distinctive phase in the production of the product, but rather exists as strife between the design engineering and manufacturing departments. A few companies, however, do recognize product redesign as an important step in production and delegate to a vice president in charge of manufacturing engineering the function of coordinating the design and production of a product so that manufacture is as economical as possible.

Product design engineers devote their entire attention to designing for ease of manufacture. Some of the most ingenious examples of product redesign can be found in the production design engineer's journal, *Product Engineering*. Articles by such men as Erwin Rausch set forth design principles aimed at simplifying manufacture and assembly methods. For these men, the idea of making possible automatic manufacture by redesigning the product is far from new. Good product design engineers are familiar with such concepts as the *doctrine of least constraint*—a systematic approach in designing component parts to allow maximum freedom in manufacture.

As a variation, let us consider what many product design engineers would consider an everyday example of product redesign for semi-automatic rather than fully automatic production. A few years ago one of the country's largest manufacturers of home ranges was producing two price lines with four style variations in each price line. All style variations were in two basic types, making sixteen variations altogether. As production runs on any one variety were limited to two or three thousand, it was not economical to build automatic production machinery. Nor was it economical to use a high-speed press, because the manufacture of sixteen dies would require very high runs on each die to justify their cost.

The company placed in charge of manufacturing engineering a high-level man to redesign the ranges so that high-speed, semi-automatic machinery could be used. Through redesign, all sixteen variations were reduced to one basic body style of a wrap-around sheet metal type. Two styles were created merely by changing the position of the heating element. Other variations were based on the control panel at the rear of the range. The size of the panel and the number and kind of controls and other features on it determined the prestige position of the range within the price line. The panels were produced on separate, smaller assembly lines and were then riveted on the finished ranges. Major distinctions between price lines were made by altering the top without changing the basic body design.

With the basic design it became economical to buy a high-compression press. Used in all sixteen variations, the press is operated continually. With need for only one die, year-to-year variations are cheaper.

After the new production line had been set up, it was necessary, in order to meet competition, to inaugurate a new line priced midway between the two lines being produced. It was feared that many of the economies of the standardization program would be negated; but, interestingly enough, all that was necessary was to vary the panel and a few of the accessories. This was easily done. Models for the new price line were produced within two days. Thus, redesign permitted semi-automatic manufacture of a product that formerly required many manual operations. At the same time it helped bring this product within the price range of many people who had not been able to afford it.

Redesign in the case of the home ranges was far less basic than in the case of Sargrove's radio circuits. It was more an instance of clever *standardization* rather than of basic *rethinking* about the product. Both elements, however, are important. There are, of course, definite limitations to the extent that we are willing to accept standardization of product for the sake of automatic production. But if certain basic parts can be standardized, with style and functional variations introduced through other parts, positive gains can be made.

RETHINKING

Automation can often be achieved only by *rethinking*. If a product or process does not lend itself to automation, perhaps it may be redesigned so that it performs the same functions in a different way—a way that *does* lend itself to automation.

Rethinking is an attitude. It is an ability to get outside of a problem that seems insoluble and approach it in a new and perhaps wholly different way. It is a constant re-examination of whether the problems we are attempting to solve are the problems we really should be trying to solve. It is asking ourselves: Should we produce this product differently? Should we try to produce a different product that will serve the same purposes? Rethinking is a constant awareness of the end functions of a product and a continual questioning of whether those functions can be performed better or equally well by a slight variation in the product or perhaps by total change to a new product that can be produced automatically.

Perhaps the best way to describe rethinking is not in terms of what it has done, as in the examples of steel castings, radio circuits, propeller blades, and kitchen ranges, but in terms of what it can do. For this purpose, let us select a process not commonly associated with automation. This will help illustrate the fact that this new technology is not limited to manufacturing, nor is rethinking a purely industrial phenomenon. We shall try to rethink the problem of the New York Stock Exchange.

THE STOCK EXCHANGE

We often become so accustomed to doing things in a certain way that we no longer question the basic purposes of our actions. This happens in all areas of human endeavor. As time goes on, we are likely to decorate obsolete processes with new gadgets and then deceive ourselves into thinking that we have made improvements. The world is rife with examples, but none is more typical than the New York Stock Exchange.

Characterized as the nerve center of American industry, the exchange is really a glaring anachronism. On the floor of the exchange as in the ancient market places, the traders stand at their posts and offer wares—not stone jugs, but stocks and bonds. Hundreds of men swarm over the paper-strewn floor. Messengers dart to and fro with scribbled bits of paper. The glitter of a few modern devices such as the high-speed ticker tape (which records what has happened but does not participate in the action) is so blinding that we never question the basic process.

How can the exchange be automatized? A faster ticker tape? Walkie-talkie radios from office to floor broker? Conveyor belts for handling the papers? These all sound workable, but they amount to no more than adding gadgets to the existing process. These gadgets may be useful in saving manpower, but they represent no basic improvement in the process itself. What is called for is something completely different from the exchange floor as it exists today.

We start by thinking of basic functions. The basic function of the exchange floor is to provide a means for interchange of information concerning the sale, and purchase of stocks and bonds.

Can this function be performed automatically? Computers are designed primarily to receive, manipulate, and communicate information. They provide a means for eliminating the exchange floor altogether.

Today, after receiving a customer's order, a broker relays it to a representative on the exchange floor by telephone or messenger. The "floor broker" completes the transaction by finding another floor broker interested in his offer. Information about the transaction is relayed back to the brokerage office as well as to the ticker tape machine operators.

With an electronic stock exchange the broker, instead of calling his man on the exchange floor, would put the order into a machine in his office. This could be accomplished by a process as simple as dialing. The office machine would be connected electrically with a central computing mechanism that would keep records of the market in each stock on magnetized drums. These are similar in principle to magnetic tape recorders except that a coded electronic impulse rather than a voice is recorded. Orders to buy and sell at various prices would be stored as received and executed in sequence of receipt as the market fluctuates. Dials for each stock would indicate on the machine in the broker's office the price of the last sale, the bid and asked prices, and the quantity being offered.

The memory unit of the computer would contain an up-to-the-second record of all orders to buy and sell stock for each company whose shares are traded on the exchange. As information can be added to or deleted from the memory unit as a regular part of the machine's operation, a complete set of orders for future trading can be recorded magnetically. This is the function now performed with ink and paper by the broker who keeps tally of orders that are to be executed at prices other than those prevailing at the time they are placed.

If an order to buy U.S. Steel at 45 is received and if Steel is selling above 45, the order is entered by the broker and is recorded by the magnetic "memory," coded as to the purchaser and the time. When someone offers Steel at 45, the transaction is automatically consummated, provided that the investor did not change his mind in the interim and that other orders to buy at 45 were not entered previously. Information about this sale is transmitted simultaneously to the brokerage office selling the stock, the brokerage office buying the stock, the "last sale" dials of all the machines in the offices of member firms, and it is also automatically reported on the ticker tape—which never gets behind the market. From there on, all brokerage activity that normally takes place continues as it does today. The firms could clear the transactions through the Stock Exchange Clearing Corporation—or this could be done continuously by machine for record purposes. But as stock certificates would have to be physically transferred and changed, an automatic clearing device would seem to add little to an electronic exchange.

Rather than making the whole process automatic, the best answer seems to lie, as it so often does, some place between the present system and a fully automatic one. Replacement of the exchange floor is entirely practical both technically and economically. Mechanization of the brokerage office and the clearing house is of dubious value.

The proposed stock exchange would produce no change in relations between investor and broker. Automatic quoting of prices, although technologically possible, would not be introduced because the customer's call to the broker is too often a request for advice and comment along with a request for quotations. The process of buying and selling stock with an electronic exchange would perhaps be a bit faster, but the execution of orders is rapid enough at present. For the investor, things would remain just as they are today. But the electronic stock exchange would alter the relationship between brokerage office and floor representative by eliminating both the representative and the floor.

The machine we describe can be built today by any of a number of capable manufacturers. Despite the fact that the cost of electronic equipment is high, a stock exchange computer could produce savings of such magnitude that the entire outlay could be recouped within two years, even allowing for reduced brokerage commissions.

No new invention is needed to build an automatic stock exchange. The problems are not technical. Rather, they are partly a lack of knowledge by financial people of what is technologically possible and partly too great a preoccupation with present practices to allow basic rethinking of the functions of the Exchange. The former difficulty, lack of knowledge, can be overcome in rather short order; the solution to the second problem is not so simple. "Mechanization" has thus far taken the form of new gadgets—faster tickers, flashing lights, and photoelectric cells—rather than replacement of the whole antiquated mechanism of the exchange floor.

But if all of this is really possible and economical, why hasn't someone built such a machine? There is no clear-cut answer. A great many practical innovations are never adopted. In the present case two separate groups of people, stock brokers and engineers, are involved. Neither is particularly aware of the problems of the other. In fact, the present operation of the Stock Exchange is scarcely considered by the brokers to be a problem.

Management consultants frequently encounter such management "blind spots" about the existence of certain problems. Recognizing that a problem exists many times seems to be harder than solving it. It is in large measure because the stock exchange floor is not perceived to be a comparatively antiquated and inefficient operation that steps have not been taken to rethink and automatize it.

UNIT OPERATIONS ANALYSIS

A systematic analysis of production processes can help greatly in providing a basis for effective rethinking. Too often production analyses consist of lists of the specific steps being used to manufacture a product. Such lists are important and useful for certain purposes, but a fundamentally different approach is needed. It would be highly useful to analyze processes in terms of the specific functional elements or units common to the manufacture of several, perhaps dissimilar, end products. In this way, process steps can be classified and studied according to the functions being performed, aside from the products resulting from the process.

An important step has been taken in this direction by George Granger Brown and a group of his associates at the University of Michigan. In *Unit Operations* * the Brown group studied a number of industrial processes to identify the operations common to all of them. These *unit operations* were then analyzed and examples given of their part in various industrial processes. A definite stand was taken by the Brown group against the tendency of specialists to accentuate the individual characteristics of operations in certain processes. It is as a result of this tendency on the part of specialists that virtually each industrial process has a different rationale and terminology.

The Brown group, primarily chemical and process engineers, focussed their attention on the process industries. The greater part of their work deals with the handling of fluids rather than the handling and processing of solids. However, their approach is good, and their classification of unit operations is an excellent illustration of a type of analysis that provides a sound basis for process design. The following is their classification, modified somewhat for presentation purposes:

    Screening
    Size reduction of solids
    Handling of solid materials
    Flow of solids through fluids
    Classification

* John Wiley & Sons, Inc., 1950.

Flotation
Sedimentation
Transportation of fluids
Flow of fluids through porous media
Centrifugation
Fluidization of solids
Separation by mass transfer
   Solid-liquid extraction
   Liquid-liquid extraction
   Vapor-liquid transfer
   Adsorption
Heat transfer
   Conduction
   Convection
   Radiation
Evaporation
Crystallization
Agitation
Mass transfer
Simultaneous heat and mass transfer

An extension of this kind of analysis to industries manufacturing discrete product units would lay the groundwork for a systematic, wide-scale approach to automation. For example, an analysis of machining processes in terms of functions common to the manufacture of different products in different industries would be of inestimable value to those trying to redesign a process in terms suitable for automatic control.

To some extent the fact that we are beginning to recognize materials handling as a separate and distinct function indicates that some thinking has been done along these lines by process and design engineers in their everyday work. Nevertheless, a systematic and thorough analysis of all manufacturing in terms of unit operations would help in providing a sound basis for effective rethinking. As George Brown states in the introduction to *Unit Operations:*

By studying the unit operations themselves and their functions the engineer is trained to recognize these functions in new industrial processes; and by applying his knowledge and skill in the corresponding unit operations he is able to design, and construct, and operate a plant for a new process with almost as much confidence as for a proved process.

We might extend rethinking to the organizational structure of the firm, also. One of the impediments to rethinking of products and processes has been that the traditional division of responsibilities has the effect of localizing the areas in which rethinking is done. Almost by definition, however, rethinking must be done on an extremely broad basis—viewing the objectives of the entire organization as a whole. It cannot be confined to the product design engineering department. It must be an attitude, a state of mind, permeating the entire organization. An organization so structured as to maximize contact, interchange, and correlation between the product design personnel and the production, sales, and other departments is, of course, desirable for reasons other than ease of automation, but it is the problems of automation with which we are concerned.

An attempt has been made to demonstrate that the extension of rethinking to everything affecting the product and the process is an essential step in making machines automatic. We are now ready to discuss the problems of applying both the new technology and the rethinking concept to the machines.

4

Making Machines Automatic

AFTER a product or process has been redesigned for automation, the problem still remains of fitting controls to the process and of producing the product, however much both process and product have been redesigned. What kind of automatic machinery shall be used?

The continuous process industries, such as oil refining, are already highly automatic. But in the fabricating industries, except for those having very long runs of nonvarying product, little attention has been given to the automatic manufacture of discrete units. Our primary concern, then, in this chapter will be with the problem of automatizing the short- and medium-run manufacture of discrete units of product. A unit manufacturer dealing with separate parts faces far different problems than does a chemical manufacturer.

SINGLE-PURPOSE MACHINES

The most obvious answer to the problem of automatic production would seem to be to build one large completely automatic machine that would perform every operation necessary to convert the raw material into the finished product. This has been done in many partially automatic factories: a Coca Cola bottling plant is perhaps the most familiar example of an almost automatic factory that handles discrete units of product.

The packaging industry provides numerous examples of the use of highly automatic single-purpose machines of which the Coca Cola bottling machine is but one. Many milk bottling plants are highly automatic; the empty containers are fed into the machine by hand, but everything else is done automatically. The canning industry provides even more spectacular examples. In some cases the entire canning process, including the manufacture of the cans and cartons, is completely automatic. Sheet metal is fed into one part of the machine, cardboard into another part, and the produce to be canned into a third. The remainder of the process is automatic: the metal is cut, rolled, and soldered; the end is put onto the can; the can is filled; the air is exhausted; the can is sealed; a label is affixed; a group of cans is packed into a cardboard container which meanwhile has been automatically assembled; and the container is sealed.

The textile industry, although not handling discrete product units, provides other examples of single-purpose, fully automatic machines. Indeed, machines of this type have made textile manufacture possible in its present form; some of the earliest automatic control devices were developed for the control of textile machinery. Automatic textile machines are highly efficient in performing the task for which they were designed (although one has the feeling that weaving threads into yard goods is not the final answer to the problem of making clothes). One operator in the United States can tend one hundred and four fully automatic cotton weaving looms and even mend broken threads, whereas one operator is required for every eight *semi-automatic* looms such as those used in Great Britain.

In the continuous process industries, we found that by altering the refining of petroleum and the manufacture of certain chemicals from batch processing (a single vat processed one step at a time) to a continuous processing of material, it was possible to build what is, in effect, a fully automatic machine. Raw material continuously flows into one end and passes through various pieces of equipment until it comes out as a continuous stream of finished product at the other end. Actually, as separate pieces of connected equipment are used, such a system has considerable flexibility. In this respect it is somewhat different from a large, single-purpose fabricating machine.

The type of thinking and designing that converted the chemical and oil industries from batch processing to continuous processing could contribute much to the automation of the discrete process industries. The conversion utilized the special characteristics of the feedback controls in designing continuous processes susceptible to automatic control.

Attacking the problem of automatic processing of discrete product units with the same frame of reference, a single, fully automatic machine seems to be the solution. If controls are available which allow the machine to correct its own mistakes and to operate according to a predetermined pattern which is built into the machine, then the most sensible way of using these controls would seem to be to build them all into one large machine. There are many people in industry who feel that, if complete automation is achieved, it will be achieved along these lines and that the answer to producing anything automatically is to work gradually toward the large, single-purpose machine.

Fully automatic, single-purpose machines, however, are suited only to a very special case in the economy, the case of an extremely large run of a nonvarying product or of an only slightly varied product. When a firm invests in a packaging machine, the assumption is made that essentially the same type of package and the same type of packaging operation will be used for a long time. Similarly, when the Coca Cola plant invests in a new bottling machine, the assumption is made that the style and the size of the Coca Cola bottle will not be greatly changed for some time.

Whatever flexibility there might be lies within a limited range of package sizes and, of course, of the content packaged—a change from orange juice to tomato juice, or a variation in the formula for Coca Cola syrup would make little or no difference in terms of the machine's operation.

Flexibility is the major factor to be weighed in determining whether the large single-purpose machine is the real answer to a more widespread use of automation in the manufacture of discrete units.

To illustrate the point better, let us consider the fully automatic, single-purpose machine used in finishing automotive cylinder head castings.

In fabricating each casting it is necessary to bore holes, ream the holes, machine flat surfaces, and thread certain holes for the insertion of valves. Until the last decade each of these operations was performed by a specialized workman on a separate machine. One man would do the milling on a milling machine. The cylinder head would then be removed from this machine and stored on a pallet. When a number of them had been finished, they would be moved to the boring machine for the next operation. Then they might be moved to a reamer, and so on.

During and since World War II, automobile and airplane manufacturers have been able to do automatically all the fabricating operations on a rough cylinder head casting without moving the casting by hand from one machine to another or operating each machine independently. This has been accomplished by building a single large machine eighty or more feet in length. This machine has a series of stations at each of which a machining operation is performed. The machining time of each operation has been equalized and synchronized so that the line moves forward uniformly. A line of cylinder castings moves on a track-like mechanism along the entire length of the machine. The entire line moves forward one station, stops while all the operations are performed simultaneously, and then moves on again.

The cylinder head machines work extremely well; they eliminate the handling operations and allow completely automatic production. They insure a more uniform product, cut down on spoilage and rejects, decrease direct labor requirements, reduce the inventory tied up between machines, and step up the rate of production.

The difficulty is that a firm buying such a machine is committed to the production of a specific type of cylinder head for a number of years. Even in the automobile industry where yearly runs are large, small variations are possible, but major changes cannot be made without scrapping a large investment—several hundred thousand dollars of machine tools in the instance of the cylinder head machine. It has been possible in some cases to make major changes in the product by adding machinery at the end of the line to make the alterations. But the line is then no longer fully automatic, and much of the economy of the original machine is lost.

Many of the automatic and partially automatic factories so widely written about depend upon inflexible production machinery of this type which is adaptable only to extremely long runs of product and useless for the far more common medium and short runs. This fact is ignored by those writers who imply that in time such single-purpose machines will produce everything we use.

It is true that the plants of today contain the elements of the automatic factory of tomorrow, but *these elements are the automatic and flexible materials-handling and machine-loading devices.* For medium and short runs, flexibility is essential. Only when the problem of automatic production of medium and short runs of product is solved will automatic control mechanisms be used to fullest advantage and on the widest scale.

NEW APPROACH TO MACHINE DESIGN

One approach to the problem of automatic production for medium- and small-run plants was presented by Eric W. Leaver and John J. Brown.* They proposed a radical change in machine design, that is, to design machines in terms of the functions to be performed rather than in terms of predetermined end products.

Present production machinery is the logical outgrowth of the eighteenth century and a philosophy of design that keeps the *product* rather than the *operation* in view. The Benthams in

* *Fortune,* November 1946.

designing machinery to mass-produce pillow blocks were primarily interested in the blocks, not in the operations that produced them. This highly practical Anglo-Saxon interest exclusively in results has led industry far from the rational line of development in production machinery. It has led to increasingly uneconomic specialization. . . .

The authors propose an entirely different view of machine design that will concentrate on basic operations rather than on the product. In this view the machine is considered as the total production unit, combining both the machine itself and the functions of the workman who might operate it. The parts of this machine will be as specialized as any of the present production tools, and may be even more complicated, but they will be specialized and complicated in a different way. This question of specialization is the fundamental difference between the design theory of the eighteenth century and that suggested here. . . .

The new machine is made up of many small units plugged together. Each unit is capable of performing one function, and several plugged together will be capable of doing all the operations required to build a given part. . . .

Specialization of machines in terms of end product requires that the machine be thrown away when the product is no longer needed. Yet the work the production machine does can be reduced to a set of basic functions—forming, holding, cutting, and so on—and these functions if correctly analyzed, can be packaged and applied to operate on a part as needed.

In many ways Leaver and Brown seem to propose something similar to the unit operations analysis of George Granger Brown and his associates at the University of Michigan. To this extent, it is a type of thinking greatly needed in this field. Moreover, in attacking the problem of automation by means other than the large single-purpose specialized machines, Leaver and Brown come close to the heart of the automation problem.

Their solution, however, is not entirely practical in such industries as metal working. They propose a series of small, functionally oriented machines that can be "plugged" together. Designing functional elements and plugging them together is entirely practical when dealing with electrical or electronic circuits, but it is a different problem to "plug" machine tool units together.

A major problem in the case of discrete product units is the necessity of moving the product unit from one "functional" tool to another. There is the added problem of varying size of product. A turning tool, for example, must be designed to turn a certain size-range of workpieces. Ideally, a Leaver and Brown turning tool would be capable of machining a large range of sizes, perhaps everything from a watch part to a mounting for gun cases. But a turning tool with so great an operational range requires a capital investment that is clearly uneconomical. The dilemma is this: If we are to have functionally oriented machines, they must obviously handle products varying in size, or we are worse off than we were with the large, single-purpose machines. On the other hand, to be practical they must be designed for a rather limited range, or we are faced with the problem of excessive investment.

If we decide, as we must, on a turning tool that handles a limited size-range of product, then we must have a *family* of turning tools for different size-ranges. In addition we must have a family of tools for each of the other machining functions such as milling, grinding, and lapping. We are then faced with the problem of *plugging* these machine tools together. This requires some type of transfer mechanism; and we need a large range of sizes of transfer mechanisms. With such an arrangement we actually lose ground because it gives us a long series of single-function machines and, in effect, requires us to hook them together.

A better solution to the problem would be, perhaps, to continue to think in terms of functions but to think of *bundles of functions,* that is, of groups of related processing functions. In many fabricating operations a clearly definable group of functions is usually performed upon a related group of products. For example, there may be a series of turning operations, followed by a drilling operation. If these could be performed at the same time in the same machine, a saving in machining time would be achieved. This would also save equipment cost, because it would not be necessary to transfer the product from one machine element to another. This comes down to something much like our present-day *production machines.*

Production machines are the midway stage between the hand-operated machines such as drill presses and lathes and the fully automatic machines common in the packaging industry. Production machines perform a *series* of machining functions semi-automatically. The *chucking machine* is an example. It is similar in many ways to a lathe except that it has a number of spindles, or heads, arranged in a circle. Each spindle holds a workpiece for machining and spins individually; but all may be revolved around the common center without changing their positional relation to one another. The machine performs a series of turning and cutting operations on each workpiece. By holding a number of workpieces—typically four, five, six, or eight—in as many heads (or chucks), it is possible to perform a series of machining operations simultaneously upon the workpieces. If the machine has six heads, for example, six different sets of operations can be performed upon each piece. The machining operations on each piece are performed in a series of stages. If more than one tool is moved against the workpiece at one station, it is possible to perform more than one machining operation on the workpiece at the same time.

After each operation is completed and the workpiece is ready to move to a new stage, the large circular head of the machine indexes, or moves one place, and the various spindles change places as do Alice, the Mad Hatter, the March Hare, and the Dormouse at the tea party. The tools that perform the operations are held at each index position so that the workpiece is moved from one set of tools to another and finally back to the starting position where it is removed from the machine and replaced by a new workpiece. Although remaining at a particular index position, the tools can move in relation to the workpiece. It is possible to take a rough cut, then a finer cut, and then a finishing cut at three separate indexing positions of the same machine. At the same time it may be possible to take another type of machining cut, perhaps a facing cut, at the same index position at which a turning cut is made. There are many other examples of production machines, such as automatic screw machines (one of the most highly automatic machines in common use in industry today) , automatic milling machines, and automatic drilling, boring, and grinding machines.

These machines, as we have seen, perform a certain group of operations upon a product within a limited range of size variations. In designing flexible machine units for an automatic factory, the starting point would be a group of functions commonly performed on a class of product. These units would be built into a machine that could handle a limited size-range of workpiece and perform its functions automatically. This solution, of course, represents a compromise between a fully automatic machine and one that performs a rather wide range of functions. But it is a compromise that lends itself economically to many of the tasks of American industry. If the automatic machine problem for medium-sized runs of product could be solved by means of automatic machines similar to our present-day production machines, we would perhaps have the beginnings of a reasonably flexible and adaptable automatic factory.

To some extent it should be possible to adapt production machines to entirely automatic operation simply by replacing the operator with an automatic control and by adding materials-handling devices. These production machines are controlled partly mechanically, partly electrically and hydraulically. The tool is usually moved by means of a hydraulic control, but the hydraulic control is actuated by an electric or electronic device, and the machine is tended (turned on and off and fed) by a human operator. Considering the large national investment in production machines, an automatic control device that could replace the operator while continuing to use the present basic mechanisms seems highly desirable; the addition of automatic controls to our present machines is the obvious and most economical course for certain types of equipment, even though we design our future machines around the *bundle of functions* concept.

But the economy of utilizing existing machine tools should not be overemphasized. More than forty-three per cent of this country's machine tools are ten years old or more. When it is realized that rapid technological changes make many machine tools virtually obsolete upon delivery, and that we spend over eight billion dollars a year in industrial maintenance, the advantages of utilizing old equipment become somewhat less desirable from the viewpoint of both the owner and the national economy. Although new machinery may require a large cash outlay and the writing off of some depreciation, it may well result in cutting a company's operating costs considerably. In addition, the total national output may be far greater as a consequence of scrapping much obsolete, though not fully depreciated, equipment and replacing it with new and more productive equipment. Too often purchasing policy is based upon the extent to which existing equipment has been written off, rather than upon the potential savings made possible by new equipment. The stifling effect of current tax policy in this regard is also a factor.

Whether existing production machines are converted or whether new types of machines are built to utilize automatic control mechanisms along the lines suggested by the *bundle of functions* concept, it is instructive to consider briefly the part played by men and the part played by machines in industry today.

To a great extent, man's function is the tending of machines. He puts the raw stock, or partially machined workpiece, into the machine and positions it. He removes the finished workpiece. He visually inspects each workpiece as it is placed into the machine and as it is removed. When trouble arises, he stops the machine, ascertains the cause, and if possible corrects it. The operator may perhaps do more than look for obvious defects; he may use an inspection device such as a micrometer or a plug gauge. In addition, he may record the number of pieces machined, rejected, and spoiled. He uses his senses of smell, sight, and hearing to determine whether the machine is operating properly. In short, he acts as an over-all control on the machine, while the machine performs the actual fabricating function.

Production machines do not index unless each operation at each stage has been completed and each workpiece has cleared each station. If something fails—for example, if a part breaks—the operator must turn off the machine and replace the part before the machine can resume operation. The operator also changes tools, although this function may be performed by another individual—a *setup* man who replaces worn-out tools and sets up the machine for each new series of operations. In many ways the operator acts as a materials-handling device and a sort of super inspection and control mechanism while the machine performs the manufacturing function.

Now, if we could couple a group of production machines, or similar machines designed around the *bundle of functions* concept, by some form of inexpensive and flexible materials-handling equipment, and add a control mechanism to do the work normally done by the operator, we would have a factory completely automatic in terms of direct operation, although there would still be need for considerable indirect labor.

This was the approach used by the Harvard research group in designing the hypothetical automatic piston factory for the manufacturing report, *Making the Automatic Factory a Reality*.* This piston factory consisted primarily of a series of existing automatic production machines connected with automatic materials-handling equipment and controlled in part by their own built-in

* Diebold, John, *et al.*, published privately at Harvard University and reprinted in part by Griffenhagen & Associates.

operating controls and, in an over-all manner, by a small digital computer which performed in many ways the same functions that the operator and the production control system now perform on these machines.

A number of problems must be solved before such an arrangement could become practical. There is the problem of replacement of worn tools. This could be handled automatically by magazine loading equipment or left to the setup men. The new sintered carbide tools can run for an entire work shift of eight hours or more. The Machinability Research Porgram, conducted by the Curtiss-Wright Corporation * under the sponsorship of the United States Air Force, found that systematic microstructure analysis of both tool metal and material being machined will substantially increase the effective wearing life of tools. Thus, the tool-change problem which once made automatic machine tool design very difficult is being alleviated.

Is it actually possible to build the flexible materials-handling equipment necessary to make an automatic factory economical for short runs of product? Can the control equipment that replaces the operator be economically produced in a flexible form? The answer given in *Making the Automatic Factory a Reality* was that these things can be done and, during the next decade, will be possible for an ever-widening area of industry.

The most important single factor in designing such equipment is to think in terms of end functions. The greatest pitfall to avoid is the assumption that the design aim is reproduction of the hand movements of the operator or laborer. To attempt such a reproduction,

* Machinability Report, Volumes 1 (1950) and 2 (1951), published by the Curtiss-Wright Corporation.

particularly for a nonproduction machine (hand-operated lathe, drill press, or milling machine), might result in the design of something very much resembling a Rube Goldberg cartoon. Such a device might be mechanically feasible, but it is not likely to be the best or most economical solution. *The strong tendency in this direction may be due to the assumption that human hands guided by human brains represent the optimum efficiency and should therefore be copied as closely as possible. Such an assumption, in most cases, is unwarranted.*

AUTOMATIC MACHINE LOADING

For an example of effective automation, let us think of the problem of loading piston castings into a chucking machine in a completely automatic piston factory. Normally an operator takes a casting from a stack and places it in the chucking mechanism in a precisely calculated position by aligning a small reference point cast on the piston head with a similar point on the head of the chucking machine.

In the hypothetical automatic piston factory in the Harvard report two reference points are cast on the head of the piston. The piston is delivered to the machine by a conveyor from which it falls onto rollers and is automatically positioned by a simple mechanism that stops its rotation when reference points are aligned. The piston is inserted by a simple ram. The device could be built quite cheaply and is only one of several workable solutions to the problem.

The machine is loaded without the necessity of reproducing hand motions. Again, as in many other instances, it is simpler to perform the same *function* in a new and different way rather than mechanically to

68

reproduce hand motions. Of course, rethinking in terms of functions is necessary.

In addition to the development of flexible, fully automatic production machines and of automatic loading mechanisms for these machines, it is necessary to find a way to move the product automatically from machine to machine. This presents an interesting and important problem. It seems evident that for fully automatic plants some type of flexible and universally adaptable materials-handling equipment will be necessary. A new conceptual scheme for thinking about the materials-handling function is needed to develop such equipment.

Again, the processing industries offer much in the form of fruitful thinking about automatic production. The concept of manufacturing as a continuous process rather than a series of separate steps is useful as a basis of analysis for any manufacturing procedure. Thinking about production in these terms has not only led to the modern oil refinery, it has also resulted in the assembly line. And, although a workpiece cannot move *slowly* through a lathe, it *is* possible to design parts of the factory other than the assembly line in terms of a continuous and automatic flow of material.

This has been done with the warehousing function at General Electric's newest television plant. The concept of warehousing as a static operation is discarded; it is instead considered to be a slow, adjustable flow. The Ford Motor Company has led the field in applying this concept to actual manufacture. At Ford the production machines are connected by automatic materials-handling equipment. The movement of many product units from one production stage to another is entirely automatic. At the DeSoto engine plant, production is also arranged in a continuous flow, with automatic movement of workpieces from one stage to another.

It would not be practical for plants with small runs of product to adopt the automatic devices used by Ford and DeSoto. It may never be possible for manufacturers with short runs of rapidly changing product to utilize automatic handling devices. But this important distinction should be made: the continuous process approach is not inexorably tied to long runs of unvarying product, although the carrying out of this approach by means of a single-purpose machine may be. The application of the continuous flow concept to the discrete unit processes, to be economical, requires *flexible* materials-handling equipment that can be easily purchased or rented rather than specially designed so that the equipment can be readily adapted to accommodate changes in the product handled.

The basis for such equipment can be found in many existing plants. But these devices are rarely thought of in connection with automatic factories because the spotlight has been on the large and spectacular pieces of special-purpose equipment and not on the simplicity, flexibility, and economy with which some companies have solved portions of their materials-handling problem.

In the hypothetical automatic piston factory the materials-handling problem was solved by means of flexible materials-handling equipment of a rather conventional conveyor type, but using devices that allowed the storage of materials between machining processes without manual unloading of conveyors. The conveyor could hold a piston casting or any other similar castings that might conceivably be machined in a plant using the same kind of production machines used for producing pistons. By rearrangement of machines and conveyors, the plant could then fabricate a different product.

The initial and conversion costs of the automatic materials-handling equipment will, in some instances, exceed the savings resulting from reduction of direct labor, smaller in-process inventory, and ability to operate more easily on a twenty-four-hour basis. Much depends upon the number of special devices that must be built for each new installation. If materials handling were to be analyzed in functional terms, as George Granger Brown and his associates have analyzed the unit operations of the process industries, there will be justification for mass-producing standard pieces of automatic materials-handling equipment that can be made available to industry cheaply. Standard conveyor equipment is already available. What is lacking is equipment for lifting, turning, and precise positioning of product units of varying size. Equipment of this sort is now custom-built.

The hypothetical automatic piston factory described earlier dealt with one shape—the short cylinder. If *flexible* automatic materials-handling equipment is really to be designed and built on a broad basis, the requirements for other product shapes must be analyzed—the long cylinder, the cube, the short rectangle, the long rectangle, the V shape, the flat sheet, the sphere, and so on. A family of machines capable of handling them must be built. When that is done, automation will be made possible in a great many areas of the economy where it is not possible today. It should be emphasized, however, that a much higher degree of automation is possible even today—both economically and technologically—than has been achieved.

The synchronization of a series of different machines is not unique to an automatic factory; it exists in any production line. But in an automatic factory the synchronization of the machines and of the machine output is critical. Discrete product units do not "flow" through a series of machine tools as petroleum flows through a series of processing units. Nevertheless, a continual and uninterrupted rate of progress must be established and, equally important, sustained. The operating times of various machines differ; machines break down. Provision must be made for these factors, to avoid clogging or stoppage of the automatic production line.

It is therefore necessary to have an inventory of product units at each stage of the processing operation to permit the line to maintain its "flow" while any part of it is temporarily shut down for minor repairs. These supplies also allow some flexibility in control of the line as a whole.

It is also necessary, in order to achieve precise synchronization, to check constantly on the rate of output of each machine or group of machines. This function will be handled ultimately by the small digital computer.

It is not necessary to wait for cheap and readily available computers. Simple, direct, interlocking controls, although they do not allow for automatic changes in rate of output, can be used to prevent the line from clogging. Their construction does not present insoluble problems. Indeed, they are already in use in a number of plants such as the automatic automobile body frame plant of the A. O. Smith Company, the engine lines of DeSoto and Ford, and in various semi-automatic production lines.

Examples of flexible inventory equipment, however, are harder to find. Ideally, such equipment should allow a free flow of units into and out of the inventory point, in accordance with the needs of the processing machines. Such equipment might consist of a simple hopper-like arrangement. If each workpiece were separately carried on a support or *shuttle*, as it is called at the Ford plant, a rack of several tiers of loaded shuttles resting on tracks would provide a reserve inventory of workpieces near each processing machine. Some workpieces, particularly the small and sturdy ones, do not need individual handling gear; a reserve or *float* of these can be kept directly on the production line between processing machines. As a unit is added at one end, a unit is used at the other, and thus the number of "extras" remains constant.

Various other methods, utilizing magazines and hoppers, have been adopted for long, continuous runs of the same product. But, as was true of materials-handling and positioning equipment, the difficult task is to devise in-process inventory equipment flexible enough for medium-sized runs where production machinery must be altered several times a year. The needs are: (1) adaptable conveyor or carrying devices that can handle a number of differently shaped products and that need not be custom-built; (2) similarly flexible hopper, bin, and magazine feed units mass-produced of light metal and adaptable to products of different sizes.

The ordnance plant at Rockford, Illinois, built by W. F. and John Barnes Company makes remarkable use of materials-handling and between-machine inventory equipment. Oddly enough, the plant was designed solely for the production of 155-mm shell casings. There was no intention of designing equipment flexible enough to produce anything else. Nevertheless, many of the Barnes Company's solutions to the problems of automation were striking in their simplicity and have resulted in flexibility of equipment to an extent unusual in an installation of this sort. Because this plant is so outstanding an example of a completely automatic factory handling discrete units of product, a description in some detail should give substance to much of what has thus far been said.

AN EXISTING AUTOMATIC FACTORY

The construction of the Rockford Ordnance Plant was begun during World War II, was about ninety per cent completed when work was stopped at the war's end, and was finished after the Korean conflict began. The author visited the plant when various portions of the line were being tested and when the machines were being tooled for actual production. The plant was therefore not in full production, but most of the individual manufacturing processes were in operation. Covering almost a million square feet of floor space, the plant in full operation will employ about one hundred and forty workers, including engineers and maintenance men.

The plant performs all operations necessary to manufacture 155-mm steel shell casings from raw steel stock. Steel is received in the form of long bars, about six inches square in cross section and about twenty-four feet long. These bars are positioned at a series of metal-cutting saws by a manually operated overhead crane. The bars are then cut into pieces about a foot long and weighing approximately 120 pounds. Thereafter, until the finished shell case is packaged, it is not again touched

74

by human hands. Except for shells removed from the line for inspection, all operations are entirely automatic.

As the steel blocks (called "billets" here, although that is not the terminology of the steel industry) are sawed from the long bars, they roll off the saw unit onto an inclined ramp made of a series of rollers. At the end of the incline the billets automatically fall onto stirrup-shaped racks. These racks, or conveyors, are the basic materials-handling units used throughout the operation. The racks are made of light pieces of formed steel, and hang from an overhead conveyor. They are uniform throughout the plant and can carry both the raw steel billet and the finished shell casing. This materials-handling equipment is a striking example of what could be possible in the way of flexibility for a range of different products; the same device could be used just as suitably for the handling of other products.

The overhead conveyors carry the billets to the two huge, circular ovens that heat the stock prior to forming. The billet is knocked from the stirrup-shaped conveyor by a simple rocker arm arrangement that is actuated when the conveyor comes in contact with the protruding arm of a trip switch. The billet slides down another inclined series of rollers and is stopped by the inventory of billets which has been allowed to collect in front of the oven door. The loading device, a pipe-like shaft about twenty feet long with a jaw at one end, looks and behaves rather like the small mechanical birds which one finds in a novelty store, pivoting on their legs and putting their heads into a glass of water, then lifting their heads again when the water has wetted their bills. The loading device moves downward; the jaw fastens on a billet; the device then moves up, and the whole mechanism is moved toward the door of the oven. The oven door is opened mechanically, the action being started and stopped by an operator at a console. The duck puts its head into the oven; places the billet on the slowly revolving floor of the oven; releases the billet; moves back; and the oven door closes. Each billet is kept in the oven for several hours during which it revolves until it is heated to about 2000° F. The red hot billets are removed from the oven by a similar "mechanical duck" device situated close to the loading mechanism.

It is curious that the mechanical loading and unloading devices are controlled by a human operator at the console. Many much more difficult control functions of the plant are performed automatically, and this operation could readily be done by an electronic mechanism using several simple sensory devices. The explanation given was: "Well, you have to have some people in the plant." This attitude is evidently not uncommon even among people building automatic factories. Oven loading and unloading are, of course, critical operations, and it is well to have as close a check on them as possible.

Actually the human operator uses judgment in unloading the oven by not letting too many hot billets pile up and cool off before the next operation. Computer circuits could do this just as well, but at the time the plant was built computer elements were not generally available and might have required an outlay far greater than the capitalized cost of the human labor. This would not be the case today.

After heating, the billets automatically are put onto a conveyor that carries them to a machine that removes the scale. The hot billets are moved by gravity to the end of an inclined roller conveyor and into a large, horizontal forging machine actuated hydraulically whose control is entirely automatic. Positioning in the machine is accomplished by mechanical stops. The machine forces the billet into a steel die roughly equivalent to the outer shape of the shell. At the same time, the inner shape of the shell casing is accurately formed by the punching head. The piercing machine, forcing the billet into an elongated shape, approximately doubles its length.

Because of the extent of reforming and because of the mass of steel involved, forming must be done in several steps. Three successive punching operations are used.

The last forming machine ejects the castings onto a small loading platform from which they are loaded automatically onto standard stirrup-shaped conveyors that carry the hot casings for a considerable distance in order to air-cool them, move through an oven for slow cooling, and finally dump them into a water bath for further cooling. At the bottom of the water tanks is a continuously moving inclined conveyor belt that moves the shells through, and then out of, the water and loads them once more onto a conveyor that carries them along the entire length of the outside wall of the plant to the machining section. The fresh air under the eaves of the factory cools them (just as some of the most modern baking plants have reverted to this age-old method of cooling).

In preparation for rough turning, several lathes face the casings and bore center guide holes. The shells then arrive by conveyor, two at a time, at the machine that rough-turns the outside of the shell case. This machine is in two parts and is like a pair of automatic chucking machines turned end upward so that the shells rotate in a vertical position. The machine is loaded automatically by a large rotary device that stands between the two turning parts. Two shells at a time are grasped in a retractable holding mechanism. When this mechanism retracts, the whole device rotates a quarter turn and the shells are held in place to be positioned by the turning machine's center points. Each machine accommodates eight shell casings at a time and performs the identical operations on each pair of shells simultaneously. While two shells are being unloaded and a new pair loaded, the other six in the machine are in different stages of rough turning. When a pair of shells has passed through all phases of the rough turning operation, they are unloaded and again dropped on a conveyor.

The turning machines as well as the central loading device are monitored by girls who preside over large consoles of signal lights and switches. The turning operation is controlled by electrical relays using hydraulic positioning devices to move both the shell and the turning tools. When anything goes wrong, the girls alter the sequence of operation or stop the machine.

In the nose-forming operation that follows, the nose of the casing is heated by electrical induction. This leaves the nose red hot and the base relatively cool. The casing then passes through a press that rough forms the nose. The shell is removed from the continuously moving conveyor belt for inspection and replaced on the belt by a simple lever. The casings undergo more turning, drilling, and threading operations all of which are performed on machines very similar to the rough turning machine. All of these machines are in pairs with a single automatic loading device between each pair. Some are monitored; others require no continuous inspection. The casings then pass through a low temperature oven to relieve stress.

The plant was designed to operate on a five-day week

and is shut down every Friday. As the annealing requires three or four hours, however, and the shells would be ruined if allowed to cool in the meantime, the last shell casing must thus be put into the oven several hours before the plant closes. As the casings push each other through the oven, it is necessary to substitute dummy casings during the last hours on Friday so that real casings will not be left in the oven when it is shut off. The real cases are diverted to a long rack about two stories high. It has a series of tracks on which the conveyors move and an automatic elevator device for moving the conveyors to different levels. It is filled gradually on Friday afternoons and emptied on Monday mornings.

This rack is close to the ideal type of in-process inventory mechanism needed for automatic production. It does not require human attention: the loading and unloading devices have the necessary automatic controls.

This in-process automatic inventory rack is one of the most interesting things at the Rockford Ordnance Plant. The larger, special-purpose machines, although spectacular and very economical for this installation, are quite common to many manufacturing processes. But the materials-handling devices and especially this loading rack, although not nearly so spectacular, are very important.

We have seen that in this highly automatic plant several manually controlled operations are used. These control points appear to offer opportunities for further automation, especially when the operator is merely responding to a signal from the console. In addition, the inspection could be substantially mechanized. The plant as a whole, however, gives us a fine example of the advanced use of flexible materials-handling equipment which provides the key to the future automatic factory of the small producer.

CENTRAL CONTROL MECHANISM

One of the most promising industrial applications of the new technology of control, not tried at Rockford or any other plant, is the replacement of individual machine tool controls by a central control mechanism for all machines in the plant. A central "brain" for the automatic factory has been part of most science fiction accounts of future industry, and the use of a high-speed computation mechanism as an over-all programming controller for synchronizing several production lines has already been mentioned. But replacement of the repetitive investment of individual machine tool controls by a central control mechanism need not wait for perfection of low-cost computers, nor is its usefulness dependent upon an automatic flow of materials from machine to machine. Grouping the control circuits of individual machine tools in a central control mechanism can produce substantial savings in many plants as they exist today without any alteration in the methods used to handle the produce units or raw materials.

At present, each of the automatic and semi-automatic production machines used in our manufacturing industries has its own control mechanism, usually next to the machine. As has been mentioned earlier, control of the various moving parts of the machine is normally accomplished by a combination of hydraulic and electrical controls. The sequence of operations is usually controlled by simple electrical relays, sometimes by electronic devices. In almost all cases the electrical devices are in use during only a small part of the actual operating time of the machine. For example, if a chucking machine has six indexing positions, the machining operation at each index position must be completed before the entire machine will index and bring the workpieces to the next step. The indexing is controlled by electrical signals to the control box.

The control box is typically a large cabinet standing next to the machine tool and contains the control circuits and signal lights for that particular machine. Each year this cabinet seems to grow larger, for as new model machines are developed, more elaborate controls are required and a greater proportion of the cost of the machine tool is accounted for through the control device. These electrical devices are capable of functioning at a very rapid rate, far more rapid than is necessary for the control of a single machine. Yet, in a typical plant of today, a series of chucking machines will stand in a row, each having the same type of control device. An investment in controls of five to ten thousand dollars per machine is repeated, again and again. In addition, a similar investment is repeated for each group of automatic machines.

The functions of the electronic devices and electric relays are very similar, regardless of the nature of the machine tools they control. Before a machine can position and allow the next operation in the sequence to begin, the operations at each prior indexing position must be complete, all tools must be clear of the workpiece, the operator's hand must be clear of the loading position, and so on. Completion of these actions is indicated by certain electrical phenomena. These phenomena are similar, regardless of what the physical action happens to be. Thus, many of the functions in the control boxes of milling machines, chucking machines, and automatic drilling, reaming, and grinding machines are very similar and are being performed much more slowly than the capacity of the electrical gear.

The question raised is: Would it not be possible to replace the individual control boxes for each machine with central control units and use the same relay for not one but a series of chucking machines—and perhaps a grinding machine too? Such a control would utilize the electronic or electric control to its fullest capacity, or at least to a much greater percentage of its capacity than is now the case. The investment would be less, and duplication of equipment would be eliminated. The hydraulic and mechanical actuating gear on the machines would, of course, be repeated but the electrical portion of the control would be far better utilized.

Such a central control device would not necessitate replacing the operators at the separate machines with mechanical loading and unloading devices. Nor would it require automatic materials-handling equipment between individual machines. The machine tools could remain in their present state of automation. This idea is not dependent upon the achievement of a higher degree of automation; it is rather a case of using control devices more efficiently and more cheaply by using them in a new way.

For example, let us consider a series of chucking machines being used in a metal working plant for the machining of castings. It is not necessary to connect the machines physically or to replace the individual machine operators in order to use a central control device. If machine No. 1 is ready to index and the various electrical impulses have entered the control circuits and actuated the proper relays, the machine will index just as it does now. It is very unlikely that machine No. 2, also connected to the same control circuits, will have to index at precisely the same instant—we are dealing here in fractions of a second. If it were necessary for the second machine to index at precisely the same instant as the first, it is possible for one machine to be delayed automatically for a fraction of a second until the control circuit has allowed the other to index. There is no problem in constructing a simple device to delay operations until a particular circuit is clear. If all machines were ready to index at the same time, the entire group would be held up at most for a second or a second and a half. This is an extremely small portion of the operating cycle of machines and would not cause undue delay or waste of machining time. The probability of all machines being at an indexing point at one time, when they are controlled, loaded, and unloaded manually, is extremely slight indeed.

With proper study and analysis, a series of standard control circuits could be devised. These circuits could be mass produced and would fill a high proportion of the individual control requirements for most types of machine tools. These control components, if readily available, would not only reduce duplicated control costs but would substantially reduce the cost of manufacturing automatic and semi-automatic machines because at present virtually all control mechanisms are handmade and custom fitted to each machine.

INDIVIDUAL MACHINE CONTROL

The nature of the control of the machine tool unit will substantially alter as advantage is taken of the new technology. The general direction of change is quite clearly toward increasing flexibility in controlling the type of the automatic operations which the machine performs. When the automatic operation of a machine tool is controlled by a cam, changes in the pattern of operation of the machine are more difficult to bring about than when controlled by electrical stimuli. As the design of stable feed-back control systems becomes more thoroughly understood, it will be possible to construct machines whose patterns of operation—not simply actuation of predetermined sequence of operations, but individual movements—can be controlled automatically by an easily altered set of instructions governing every aspect of the machine's operation.

The problem of automatically varying a machine's operation—as contrasted with merely setting in motion a predetermined sequence of operations—has been attacked in two ways: (1) in terms of providing instructions for the machine tool by punched paper or magnetic tape and of translating these instructions into movements of the various parts of the machine by means of servomechanisms or small motors; and (2) by developing copying mechanisms which actuate the machine tool controls through a sensory device that follows the contours of a finished workpiece or of a model.

Contour-milling machines and copying lathes already use the latter approach, and a recent announcement of the Raytheon Manufacturing Company indicates that a new electronic sensing, or following, mechanism has been developed. But most present work on machine tool controls seems to be in terms of *taped* instructions. If the problem of taping—that is, of setting the instructions up in such a manner that they are useful to the machine—can be solved in such a way that taping becomes a low-cost operation, it may be possible to work out the design specifications of the product in terms of the machine tape, thus by-passing many of today's engineering drawings and operating schedules. This could mean much to the manufacturer of very short runs of product as well as to the mass producer. A number of people are at present working on this problem of designing and producing automatic controls for existing machine tools.

Some time ago the Arma Corporation, in New York, demonstrated an automatic lathe controlled by a punched paper tape. In four minutes this lathe machined a workpiece that required thirty minutes when made by an engine lathe operated by a skilled machinist referring to drawings. Tolerances were held to 0.0003 of an inch. Only fifteen minutes were required to punch the tape. This last is perhaps the more significant achievement. Unfortunately, this control has not been produced commercially because the attention of the firm has been directed to the further development of gun-directing devices.

Herman Cussins, a New York physicist, has devised a beautifully simple electrical, rather than electronic, machine tool control based on a punched paper tape using several channels of punched holes and on a series of synchronized motors that move the machine parts themselves. The Cussins device, the experimental unit of which was built by the Dacco Corporation of Brooklyn and which will be manufactured and sold by Olin Industries, can be applied to any of a number of standard machine tools. The hand controls normally used by the operator are simply fitted with small servo motors. If, for example, the cross feed of a lathe must be moved 0.0015 of an inch, instructions to move the feed this distance is given to the servo motor by means of the punched paper tape. Feed-back allows the servo to move the cross feed just the right distance, despite the fact that the metal may be harder to cut than was expected and additional time or a greater number of cutting passes may be needed to move the cutting tool the specified distance.

A numerically controlled milling machine, utilizing a digital to analogue converter, has recently been developed at M.I.T. under an Air Force contract. The biggest drawback to this control in its present form is its complexity and resultant high cost. However, it will fill the needs of certain producers, for example, aircraft manufacturers who make many similar parts, but not all at one time. It is to be hoped that cheaper versions of a similar control will soon be available. Some manufacturers are working on just such devices.

Shortly before World War II, Robert Travers, in his Control Laboratory at Worcester, Massachusetts, devised a means of adjusting for tool wear by constantly altering the angle at which a tool bit cut metal. Travers' control system is similar in many ways to the Cussins system and depends upon a punched paper tape actuating a series of small electric motors. Travers solved the tool position problem by means of a tiny photoelectric cell and a light beam reflecting from the working face of the tool. The tool holder is capable of constant adjustment by control devices. The photoelectric cell keeps the tool working always at the optimum cutting angle. Travers' greatest difficulty at first was with the development of servomechanisms to move the tools. He found it necessary to design an entirely new type of motor that could stop within a very short angle of turn so that the necessary precision could be achieved. Other grave problems arose in connection with movement of the cutting tool against the workpiece. To prevent tool "chatter" and other arbitrary movements, feed-back loop-control systems are necessary.

There are serious engineering problems in designing and building these mechanisms so that they will operate under actual factory conditions; vibration, splashing lubricant, and close tolerance requirements are some examples. As with most engineering problems there are several solutions, some economical and some not. The extensive work that has been done by men such as Professor Brown of M.I.T. in the design and building of stable feed-back systems provides a solution to the basic problem. Solving the problems of application is principally a matter of time and engineering know-how.

Another operating problem is "backlash," or gear wear, in machine tool controls. After one or two years of operation gear wear usually develops and then steadily increases. A machinist is able to take into account the peculiarities of the individual machine; but with paper-tape operation measures must be taken to insure that control is in terms of the actual change in position of the cutting tool (or other machine part) rather than the expected change as a result of a given number of revolutions of the control screw.

The General Electric Company in its Record Play-back Control has overcome the gear wear problem by recording magnetically the actual motions of the machine tool when operated by a skilled machinist. A workman produced a prototype of the desired end product by normal operation of the machine tool. While he adjusts the machine and produces the product, all of his actions are recorded in terms of magnetic impulses on the various channels of the magnetic tape. When he has completed the machining operation a new workpiece is placed in the machine and the tape is played back to the machine. All the movements of the craftsman are repeated exactly by the machine tool by means of servomechanisms moving the hand controls. They reproduce even the skill of the craftsman—the changes in speed and even the unconscious adjustments he makes to compensate for gear wear of the particular machine. The individual differences between machines of course require that a separate tape be made for each machine being controlled, and from this standpoint the gear wear problem is not ideally solved. But Record Playback Control is a very good solution to the problem of the short run of product. It is no small achievement, because it means that automatic operation of machine tools is possible for the job shop—normally the last place in which anyone would expect even partial automation.

A NEW FAMILY OF MACHINES

Nevertheless, the copying of hand motions and the automatic control of machines designed for manual operation cannot be regarded as the most fruitful use of the technology of control.

Typically, the introduction of any new technology, such as feed-back, first brings an attempt to apply this concept to the present way of doing things. The present efforts to apply feed-back to industry are directed toward adapting it to existing machinery and processes. But the new technology of control in terms of actual rather than expected performance makes it possible to build entirely new types of production machines—machines which still perform a bundle of functions but are quite different from those we now have.

Feed-back provides industry with a very powerful new tool for the design of machinery. A man has two hands, two eyes, and ten fingers with only limited dexterity. He can control only a limited number of variables at one time. All too many of our present machines have been built around these limitations. In some cases this has hampered design greatly and has led us into inefficient ways of production. With intelligent use of the feed-back principle and of the automatic control devices which this principle makes possible, we should be able to achieve entirely new types of automatically controlled machinery. This has a far greater significance than simply fitting automatic controls to our present machines. It is difficult to foresee what forms these new families of machinery will take, but we do know the areas of industry in which mechanization has been least successful—machine setup, materials handling, product inspection, and assembly. It is clear that the new technology has much to offer toward the automation of all these tasks as well as toward the automation of the office. It is equally clear that industrial automation will not be complete until all of these functions have been made **automatic.**

5

Automatic Handling of
Information

WE HAVE discussed automation of the plant.
What about automation of the office? It is probably
there that the most immediate, widespread, and fruitful application of the new technology will be made.

The primary function of the office is the handling
of information. Our discussion of the Stock Exchange
has indicated the need for and the fruitfulness of rethinking the "information handling" function.

In the plant, the materials-handling problems are
formidable obstacles to effective automation. In the
office, materials handling is the basis for use of the new
technology. We may deceive ourselves into thinking
that the materials handled in the office are papers and
cards, but actually the basic material being handled is
information. The new technology was developed primarily for the effective and rapid handling of information.

The problems encountered in one-wire transmission
of information—measurement of quantity of information, determination of patterns by which information

can be transmitted, "noise" (or interference)—are the problems which led to communications theory in its present form. It is in this area that a great amount of research is progressing at the highest engineering, electrical, and mathematical levels. Although the existing computers are designed to handle the information of the position of aircraft and problems in mathematics, these devices are readily adaptable to handling the information of business; and the information functions of business, in turn, provide a fertile field for automation.

As business grows and becomes more complex, the need for detailed, up-to-the-minute, accurate information increases enormously. Better methods of production control and market analysis, a growing body of government regulations, complex pay-roll deductions—all these are placing an increasing burden on office procedures. Although we have developed some extraordinary machines for handling information, between 1920 and 1950 there was a 53 per cent increase in the number of factory workers as against a 150 per cent increase in the number of office workers. In addition, the office has moved increasingly farther away from the manufacturing process, both physically and functionally, although there are now signs of a back-to-the-plant movement.

This chapter might have been entitled "The Automatic Office." Although the temptation was great, it was felt that this phrase might distract attention from the fact that the new technology will ultimately bring many of the office functions into closer contact with the production functions. In manufacturing firms, the function of the office is fundamentally related to the function of the plant. Recognition of this fact is of considerable importance to a fruitful analysis of office methods. When this basic relationship is overlooked, the introduction of new machinery often perpetuates eixsting procedures instead of eliminating or replacing them.

We have said that the basic form of the high-speed digital computer is ideally suited to the automatic handling of business information. Indeed, the components of the computer—the input and output sections, the central control element, the arithmetic and the memory units—are all counterparts of the functions used in the manipulation of business problems.

For example, a man analyzing a sales record will use as tools a record of past sales, a work sheet or scratch pad, and perhaps a desk calculator. All of these elements, including the human functions of control and programming, are duplicated in the computer. The arithmetic element of the computer replaces the desk calculator as well as the comparison of data by the man. The memory element retains the sales information which the man retains on sheets of paper. The intermediate data, which must be set down by the man as the analysis progresses, are retained in the memory circuits of the computer and "erased" when no longer needed. The outlining of the successive steps that the analysis is to follow, determined either mentally by the man or by an instruction sheet, or by a combination of the two, is determined in the computer by the control or programming elements.

As was explained in Chapter 2, the computer performs only very simple basic operations, but when properly programmed these individual simple operations can solve highly complex problems. It is, of course, the incredible speed at which the computers perform the basic unit tasks that makes them of such great potential value to business.

Most writing on the subject of applying computers to the handling of business information has emphasized speed and accuracy as compared with present methods. But the new technology can do far more than improve speed and accuracy.

LINKING PLANT AND OFFICE

With proper analysis and development, the control circuits that are used to operate the production machinery can be directly linked to the equipment the function of which is to schedule production. Thus the routine supplying of production data by hand, the recopying of this information, and its transfer to punched cards can all be eliminated. With proper storage systems the information-handling equipment can retain all of these data in its own storage system, printing only the portion that management must have to make decisions and retaining the raw data for automatic comparison with, for example, production planning schedules for future operations. In addition to eliminating many clerical operations, data can be provided for management more quickly and in far more useful form than at present. To accomplish this, however, special analysis of the particular information functions of each firm is necessary.

Similar analysis of the information functions of firms not engaged in manufacturing can lead to extensive improvements and economies. It is the purpose of this chapter to suggest, in a basic way, the kind of analysis that is necessary. But it is perhaps best to give first a few examples of the diverse ways in which the new technological developments can be used, in order to get an idea of the results such an analysis can produce and of the equipment involved.

93

An example of a firm which can make good use of automation but which is not engaged in manufacture is an insurance company. A large life insurance company will use ten or fifteen stories of its skyscraper headquarters merely to keep records of insurance policy information. These are usually in filing cabinets distributed evenly throughout the floors. File clerks are stationed every fifty feet and are equipped with headphones connected to the intercommunication system. When an employee needs information about a policy, he dials the clerk in charge of the appropriate section, tells her the policy number, and waits until she walks to the filing cabinet containing the policy and reads him the information.

By the use of magnetic tapes, the storage space for the policy information of a large insurance company can be reduced from ten or fifteen floors of files to 350 or 400 spools of magnetic tape which, with control gear, would occupy one medium-sized room. This magnetic tape could be kept in a low rental area far from central headquarters.

Under such an arrangement each employee who requires policy information would have on his desk a dial, similar to a telephone dial, and a small ticker tape printer, rather like the Dow Jones stock ticker but far more compact. He would dial a code number (which might even be the policy number) and thereby activate a series of relays; the policy information spool in the file department would spin very rapidly and stop at the part containing the coded information on the desired policy; the information on the spool would be "read" by a decoding device and printed on the ticker tape on the caller's desk—all within a minute. With a more complex control mechanism it would even be possible to obtain just a portion of the information in each policy.

At night the entire spool mechanism could be switched over to a billing machine operation, and all the bills and checks could be made out automatically from the same mechanism that provides the policy information. In the morning the bills and checks in their stamped envelopes would be ready for mailing, and the policy records would be again available for reference.

There are obviously many problems in instituting such a system. One is the problem of leaving space on the tape between policies for the insertion of new information. Another is altering old information. These are not, however, insurmountable obstacles and have been overcome in memory circuits used for other purposes. More serious is the legal requirement of maintaining some written form of policy record. Of course, this requirement could change in time as the competence of machines becomes evident, and as such simple expedients as duplicate recording of the information help reduce the chance of error considerably. Typically, the errors made by such machines, when properly set up, are far fewer than those made in records prepared by hand. The original policy application and correspondence can, of course, be kept to fulfill legal requirements; the magnetic record can be used for actual operating purposes. However, the original saving in floor space is decreased by retaining these documents. But some saving is still possible by using "dead" storage space; that is, storage cabinets can be stacked to the ceiling. In addition, the files can be stored in a low rental district.

Let us take another example—transportation. In this industry a great problem is the handling of reservation information. The paper work in the transportation industries is so great that General Georges F. Doriot once quipped, "The railroads are basically a materials-handling industry. They become confused, however, as to what materials they are handling and spend as much time moving bits of paper about as they do moving their trains!"

Recent attempts to rectify this situation provide interesting examples of the application of some of the new information-handling devices to business. International Standard Trading Corporation, an associate of the International Telephone and Telegraph Company, has designed, built, and installed for the Pennsylvania Railroad at Pennsylvania Station in New York an automatic reservations-handling system called *Intelex*. The following description of *Intelex* has been extracted from a speech made by Mr. J. S. Jammer, Vice President of the International Standard Trading Corporation. These excerpts are interesting, not only because the system has actually been built, but also because they describe, in some detail, the kind of problems that must be solved in automatizing information-handling processes.*

. . . The classic pose of the ticket seller, balanced on one foot with the telephone pressed to his ear as he faces an impatient passenger line while waiting for somebody to answer his call, is one that everybody wants to eliminate. Why does a ticket seller have to wait? He waits because the telephone lines are busy, and they

---

* I am indebted to Mr. Jammer and to Mr. M. A. Sunstrom, President of the International Standard Trading Corporation, for their kind permission to include these extracts from Mr. Jammer's speech before the Erie Section of the American Institute of Electrical Engineers.

are busy because other ticket sellers are calling the harassed attendants (who have the space diagram for each railroad car) at the same time. Once the waiting starts the ticket seller is not the only one who waits. The customer at the counter waits. The customers behind the one at the counter wait. The public trying to call in from the outside also waits. As a matter of fact, everyone and everything waits under these conditions except the trains that are set to go out on time, perhaps without the passengers who have been trying to buy their tickets. . . .

It became evident early in the investigation that the three biggest time-consuming operations were:

1. The time required by the clerk at the ticket window to find out what space was available.

2. The time needed by the attendant to locate the appropriate diagram.

3. The time lost in relocating the diagram when the customer who has made his reservation by telephone appears at a ticket window to buy his ticket.

Consequently, the Intelex System has mechanized these operations so that the ticket seller can determine automatically the space available within a matter of a few seconds and without reference to the diagrams; secondly, the diagrams are automatically selected from the files and laid before the attendant when an entry needs to be made; and thirdly, information regarding the customer who will call for the reservation he had made by telephone is automatically supplied to any ticket office designated by the customer.

These three elements are called Automatic Availability, Automatic Space Control and Telephoned Ticket Availability. . . .

The very essence of Automatic Availability is that a ticket seller or a telephone reservation clerk, instead of listening to the voice of an attendant while the attendant searches the diagrams, now listens to a voice record of exactly the same information, which has been prepared in advance of his inquiry, and he selects the record automatically without disturbing the attendant. The attendant now has no verbal contact with the ticket sellers or with the public, and has only one simple job to do: that is, to enter in the car diagrams the reservations that have been made. Again, while formerly only one ticket seller could talk to one attendant looking at the diagrams of a particular train, now any number of ticket sellers can listen to the same record simultaneously. Thus, although there is an automatic telephone switching system in use for the ticket seller to select the record, it is an automatic telephone switching system without a busy signal.

The second main Intelex feature is Automatic Space Control. The ticket seller, having determined from the availability record the space that will be suitable for the customer, and having pushed appropriate buttons on a newly developed device, similar to a key box and known as an order turret, before him during his conversation with the passenger, and having changed the buttons as many times as the passenger changes his mind, has only to press a "send" button. This automatically assigns a serial number which, among other things, identifies the sender of the transaction, and sends a code message to the central electronic switching equipment. Here the message electronically selects an automatic file that lays before the attendant the car diagram containing the space or spaces requested. The message appears in printed form at the same time before the ticket seller and the attendant at the space file. The attendant then enters the serial number of the transaction on the diagram and types the identification of the allocated space on the printer, which adds it to the message. Automatically, the space assignment is printed in front of the selling agent. At the same time the diagram is automatically restored to the file.

The attendant at the space file, who handles the entries on the diagrams, no longer answers the public requests for reservations over the telephone. The telephoned reservations are received by a reservation clerk who carries out the operations that have been described for the ticket seller behind the counter. However, in the case of a reservation requested by the public over the telephone, the telephone reservation clerk asks the customer his name and where he will pick up his ticket. The telephone reservation clerk, having received the space assignment from the attendant, adds to this information the data regarding the customer's name, where the ticket will be picked up, the time until which it will be held, etc. This information is then sent by printer to the sales office designated by the customer. At that point the message is put in an alphabetical file for ready reference when the customer arrives, or, if desired, the ticket can be prepared in anticipation of the customer's visit. This last, in brief, is the Telephone Ticket Availability feature of Intelex. . . .

The third point which has been mentioned, that of Telephoned

Ticket Availability, is very simple and straightforward, consisting of sending to the ticket office selected by the customer the information concerning his reservation. Where the customer is well-known, tickets can be made out in advance of his calling for them. This is a feature that could have been introduced at any time, but there were difficulties that were not easy of solution until Intelex solved the other problems of recorded availability and space control. Probably the main point of interest to traffic men is that the time limits on reservations, which were hitherto always retained on the diagrams in the bureau, are now held and operated by the various ticket offices around the city. It was thought that this might lead to some difficulty in controlling the time limits, but practice has shown that this is not the case. Indeed, the time limits are very much more easily checked from the alphabetical files in the ticket office than was the case when they were hidden among all the thousands of diagrams that exist in a bureau. . . .

. . . The Intelex Automatic Reservation System is a practical solution to the need for speed, accuracy, and economy in handling reservations. It uses only tried and proved techniques. It makes no major change in the buying habits of the public. It eliminates handling car diagrams when availability information is required. There are no telephone communications between the ticket sellers and the distributors. Permanent printed records of all transactions are automatically provided. Resale of space cancelled at the last moment is facilitated. Round-trip reservations can be made with the same facility as one-way.

The problem of the automatic handling of reservations has been attacked in quite a different manner by American Airlines, Inc., whose Charles Amman, in collaboration with the Teleregister Corporation, a subsidiary of Western Union, designed and built a small-scale computing mechanism that records reservation information in code on a rapidly rotating magnetic drum. The function is very similar to that performed by the *Intelex* system, except that fewer variables are at present dealt with by the American Airlines computer than were necessary for the Pennsylvania Railroad reservations system, with its complexity of space accommodations and variety of cars making up an individual train.

The American Airlines system, called the *Reservisor*, is installed at La Guardia Field and is at present limited in operation to flights originating in New York. A clerk desiring information on New York flights is able to ascertain space availability on any flight originating in New York during the following ten days by using a key system to send his inquiry to the magnetic drum. If the requested space (one or more seats) is available, a light flashes on the clerk's desk within a matter of seconds. If the customer decides to make the reservation, the clerk informs the computer of this fact by using of the key system, and the number of available seats is decreased by the number of seats sold to the customer. A cancellation can be indicated to the magnetic drum in a similar manner.

CONTRAST IN APPROACH

These two solutions to the problem of automatic handling of reservations (the Pennsylvania Railroad *Intelex* and the American Airlines *Reservisor*) present an interesting contrast. The Intelex system is essentially an application of electronic devices to the present way of handling reservations. It is proving economical and probably will be installed in other cities. There is, however, a question whether this is the most fruitful way of applying the new technology. Although the American Airline Reservisor required a considerable expenditure for research and development, the magnetic drum computer replaced a roomful of clerks and telephones with a small computing mechanism. The resultant saving is greater than the development cost, and present plans are to expand the system so that accounting data, in addition to reservation data, can be handled by the system. It would seem that the Reservisor is a more effective utilization of the new technology.

Design and programming difficulties lie in the way of translating the railroad problem into a fully automatic system. It is argued by Mr. Jammer that the increased cost of further automation by the *Intelex* approach outweighs potential savings. This is often the case, of course. Making a process eighty or ninety per cent automatic may produce great savings; attempting to automatize the remaining ten or twenty per cent may make the entire operation uneconomical. This indicates that some *rethinking* is needed. As has been pointed out, one hundred per cent automation may be impossible in some present procedures; however, automatic performance of the *same function* in a *different way* may be entirely possible. For example, if television tubes were to be used to view the car diagrams (mentioned as a possible further step in automizing the Intelex system), the expense of this marginal automation could make the entire system too costly. Alternatively, translation of the car diagrams into code is both possible and cheaper.

ACCOUNTING SYSTEMS

One of the best examples of automation of what once was an enormously complicated paper-work operation is the Automatic Message Accounting System which will eventually permit dialing of long-distance telephone calls throughout the country. Developed by Bell Telephone Laboratories, the A.M.A. System uses a number of standard electronic units combined into a complicated, effective, entirely automatic device. It records both local and long-distance calls, assigns these calls to the proper subscriber's account, and automatically compiles and prints the telephone bills. The device, in connection with the dialing system, will eliminate the need for long-distance operators and for their clerical work. Formerly, each time a long-distance call was placed, a notation was made by the operator for the accounting office. The Automatic Message Accounting System puts this information directly onto punched cards.

The A.M.A. System is especially interesting from the standpoint of the occurrence of errors in automatic information-handling equipment. The telephone accounting system involves many hundreds of thousands of individual notations and compilations of these notations. This is an information-handling function of such complexity that the cynics were quick to predict that if the system were operated automatically the errors would be enormous. In actual operation, however, something on the order of one error in 36,000 notations has occurred. The incidence of error is, in fact, so small that it gives great hope as to the extent to which machines will be able to take over the intricate accounting and information-handling functions of present-day business.

The examples cited thus far have been automatic information-handling systems requiring the construction of special machinery; on the whole they have been unique projects, being built primarily as experimental prototypes of what are later to become more extensive installations. However, office machine manufacturers and a number of new, small, independent electronic manufacturing companies have done considerable research and development work on mass-produced small computing machines suitable for office use. The most notable examples of machines of this type already on the market are those manufactured by International

Business Machines Corporation and Remington Rand, Inc.

In 1948, IBM introduced its 604 electronic calculator. This unit consisted of a small digital computer and control unit that could be connected to IBM card-punch machines. In 1949, the 604 was hooked to an electrical relay storage unit and could be connected to IBM accounting machinery. The assembled unit was called the CPC, or Card Programmed Calculator. These units are now in line production at IBM and are in use by a number of manufacturing, financial, utility, and transportation companies, and several government bureaus.

Remington Rand's 406–2 electronic calculator is that company's counterpart of IBM's 604. On the whole it performs the same functions in a slightly different way. Both represent attempts to use small computing machines as control units for punched card office machinery. The computers provide automatic programming for the office machinery and speed up many normal accounting procedures. For example, the IBM 604 can program from twenty to sixty calculating steps, the Remington Rand about forty. A detailed pay roll complete with all deductions can be prepared rapidly and with much less effort than with manually controlled punched card accounting. The faster operation of these units permits the use of punched card accounting equipment for new purposes.

Both the IBM and the Remington Rand calculators are at present being used for the rapid handling of income tax returns, the compilation of pay rolls, determination of stock balances, sales and price records, and many other complex accounting procedures. When properly used, this equipment has speeded accounting operations and in many cases has halved preparation time for financial statements, thereby decreasing the lag in management's knowledge of current operations.

The IBM and Remington Rand equipment is also currently used by a number of public utility companies for billing. The meter reader carries punched cards and, after reading each meter, marks the card with a graphite pencil. These cards are run through an IBM mark-sensing unit that translates the marks into punched holes which permit the use of the cards in punched card accounting machinery. These cards are then compared automatically with the customer's account, and the new information is recorded on the account card. In addition, the monthly or quarterly bills are prepared and printed automatically by the IBM equipment.

In the future, meter readers may be eliminated altogether by direct electronic monitoring from the central power station. Individual meters could be eliminated, and power consumption determined directly at a central point. The information could be translated automatically into bills and permanent records without human intervention.

Both computers illustrate the way in which flexible functional units—the computer, the electric relay storage units, the individual punched card machines for bookkeeping, accounting, sorting, duplicate-punching, checking, and mark-sensing—can be mass-produced yet individually adapted to the special and varying needs of users. Individual analysis of the user's problems determines the way the proper electronic units should be connected. This is, of course, very similar to what Leaver and Brown postulated for production machinery. In the case of office machinery, however, it is far more feasible and is being done by the most progressive office machinery manufacturers.

One of the most interesting developments in this field is the great number of small firms that have begun manufacturing computers during the last few years. Typically, these concerns have been started by engineers who worked on the large "scientific" computers. Having seen the immense potential industrial market for such equipment, they have formed companies for building small computing mechanisms that could be used by business. In some cases these firms have secured government contracts for their initial work and perhaps for the production of some of their early models. A few of the more successful have recently been bought in their entirety by large and established concerns eager to get started in this field and lacking the kind of engineers that are needed.

Typical of the companies currently offering computing equipment suitable for business use are: Engineering Research Associates of St. Paul, Minnesota, and Arlington, Virginia (recently acquired by Remington Rand), who have designed and manufactured a general-purpose, high-speed digital computer (ERA 1101); Tecnitrol Engineering Company of Philadelphia, Pennsylvania, specializing in digital computer components and measuring instruments; Computer Research Corporation of Hawthorne, California (now controlled by the National Cash Register Company), producers of computer components and several models of a general-purpose computer (notably CRC 102); Potter Instrument Company, Inc., of Great Neck, New York, marketing an entire line of high-speed electronic computing, timing, and control instruments; and the deFlorez Company of New York City, which designs and supervises the construction of computers and control systems.

The Northrup Aircraft Corporation of Hawthorne, California, has produced a desk-sized Digital Differential Analyzer (MADDIDA), as has the Hogan Laboratories Corporation of New York City (the Circle Computer). Both of these instruments are being offered primarily for use in computation problems and are suitable for the production control and other industrial scheduling.

Small, high-speed analogue computers are currently available from Reeves Instrument Corporation of New York City (REAC—Reeves Electronic Analog Computer); The Computer Corporation of America of New York City (IDA—Integro-Differential Analyzer); Beckman Instruments, Incorporated of South Pasadena, California (EASE—Electronic Analog Simulating Equipment-Computer); and George A. Philbrick Researches Corporation of Boston.

Automatic printing, typing, graphing, and other *output* devices for use with computer equipment are manufactured by the Benson-Lehner Corporation of West Los Angeles, California; the Clary Multiplier Corporation of San Gabriel, California; and the above-mentioned Potter Instrument Company.

This is by no means a complete list of computer manufacturers. In addition to those named, every major producer of office machines and a number of manufacturers of electric and electronic equipment are either currently offering, or soon expect to offer, high-speed computing equipment of a type suitable for business use. However, the fact that so many small companies, many started just for this purpose, are already marketing equipment of this type (requiring extensive research and development) whereas the giants in the field are in many cases only now beginning to overcome their skepticism, is an interesting comment on how dynamic our economy still is.

One objection which the small electronic computer manufacturers make to the IBM and Remington Rand installations is that these latter rely too heavily upon punched card machinery. It is said that the large companies have heavy investments in punched card equipment; nearly all of it is rented, and ownership is retained by the manufacturer. These companies are therefore said to be reluctant to introduce units that store information in the form of magnetic tape or drums. IBM and Remington Rand reply that magnetic tape storage is not yet fully developed and that punched cards are still far more useful at this stage of development. There is certainly no doubt about the fact that both these companies are spending substantial sums of money in the further perfection of magnetic storage.

Although future equipment will make more use of magnetic tape storage than is done today, punched cards will still be used in many instances. The objection to punched cards is that they limit the speed of a machine's operation. Computing circuits operate at enormously high speeds. The necessity of sorting a group of punched cards so that a machine can find a specific bit of information requires complete stoppage of the operation of the machine for a period of time. By storing information on magnetic tape or a continuously rotating drum, it is possible for a machine to find the bit of information much more rapidly.

There are many installations, such as IBM's Selective-Sequence Calculator in New York, that use a combination of punched card, electric relay and punched paper tape storage. In the same way, machines can be built to use a combination of punched cards and magnetic tape. Punched cards can also be employed to introduce the information into an intermediate unit. The information can be translated from the punched card to a magnetic tape on a machine entirely separate from the computer. The wire tape can then be fed at a rapid speed into the computer. A similar method can be used in the output from the computer. Since the computers operate at a speed too fast for direct printing, a magnetic tape unit can serve as an intermediary between the computer and the printing mechanism.

ANALYZING OFFICE PROCEDURES

Before computing machinery or any of the computer techniques can be applied effectively to the information-handling tasks of business, present office procedures must be analyzed. Every action and every decision must be reduced to a series of simple, logical steps which, when properly coded, will be meaningful to the computer. When stated in such terms, even the most elementary office routine can fill pages of programming forms. The more detailed the original programming analysis, and the more alternatives permitted in the original programming (or taping), the less complicated is the task of introducing new data—the need to supplement individual data with instructions for special treatment will thus be decreased.* Once the routine

* Those interested in pursuing the problem will find ample literature on the subject—Edmund C. Berkeley has prepared an excellent computer bibliography (*Giant Brains,* John Wiley & Sons, 1949). In

is analyzed and properly structured, the same machine instructions can be applied to innumerable sets of new data for recurring problems. Programs can also be devised to make the coding of such data quite mechanical so that mathematicians and computer experts are not needed in day-to-day operation of the equipment. Nevertheless, the original procedures analysis, form design, and compilation of instructions necessary for each variation in procedure present real problems to firms wanting to install computing machinery—and to those wanting to sell such equipment.

It should be clear that a businessman cannot order the necessary equipment, plug it in, and have an automatic accounting system, even when standard units are mass produced. Original installation of an electronic information-handling system is best done by an outside consulting group provided either by the firm supplying the equipment or by an independent consultant having no vested interest in a specific type of equipment. The day-to-day variations can be handled by a staff procedures analyst, trained in computer techniques. The training of such analysts in the necessary skills does not present formidable difficulties. The Remington Rand Corporation is already conducting a regular seminar in the operation and programming of its UNIVAC. IBM schools are well established for the training of personnel in the use of IBM equipment. Similar training programs will undoubtedly be started by the equipment manufacturers as more computing machinery designed for business use enters the market.

---

addition, Remington Rand, whose UNIVAC (Universal Automatic Computer) is designed specifically for business use, has prepared a two volume *Programming Manual* for use in converting the clerical problems of business into terms meaningful to UNIVAC. In addition, Remington Rand offers a seminar in the business use of UNIVAC.

The need for analysis of office procedures as a basis for computer programming has been recognized by all who have given serious thought to the business use of computing mechanisms. Attention has already been given to the systematic solution of programming problems. But there has yet been little recognition of the fact that the computer and the computer techniques make possible an entirely new approach to many of the information-handling problems of business. It is just as important in the office as in the shop to avoid the mistake of decorating obsolete processes with new gadgets. However economical the new gadgets may be in releasing manpower, maximum benefits will not be realized until we recognize that this technology presents us with an essentially new medium and until we learn to make effective use of it.

UNIT OPERATIONS ANALYSIS?

Our clerical procedures have been designed largely in terms of human limitations. Humans can easily handle only a limited number of variables, and they can solve problems of only a given magnitude with speed and accuracy. Useful information concerning business operations often calls for too great an expenditure of human effort or cannot be analyzed by humans quickly enough to be used.

What the new technology offers is an escape from designing in terms of the limitations of human operators. And it is only when we learn to organize our information-handling procedures to take full advantage of the freedom offered by the new medium that we will receive the true benefits of computer technology.

An analysis of the information function in business must be far more penetrating than the usual *procedures analysis* that now accompanies installation of office equipment. Procedures analysis resulted in a substantial portion of the over-all savings associated with the introduction of punched card equipment. However, what is required is something similar to George Granger Brown's *unit operations* analysis which structures processes in terms of functional units common to different operations rather than in terms of the manner in which the processes are carried out. The question to be asked is not: How can we handle these forms more efficiently? but rather: How do we use the information contained on these forms? Why is it gathered in this manner? How does it relate to other information we need for operation of the firm? In what ways could these data be presented to management in a more meaningful form?

To use the new technology as a speedier means of preparing the same reports that are now prepared and to treat their contents in the same way they are now treated would be a great mistake. What the new tools offer is, in many cases, *an entirely new way of handling business information.*

Perhaps this can best be explained by illustration. Let us consider a hypothetical factory in which the production processes are fully automatic. To have an automatic factory we do not require even one new basic scientific innovation. Remember also that a plant does not have to be entirely automatic in order to use automatic information-handling and -processing machinery of the computer type, or of a less complicated type, although the computer equipment will produce the greatest savings in a plant where the information can be handled continuously and automatically.

If information can be gathered as a by-product of normal manufacturing functions and operations, a great saving can be made. For this to be possible it is not necessary that the production machines be physically connected by means of materials-handling equipment. Automatic counting and inspection units and even manually operated information "input" signals would be enough. Therefore, although the description that follows is of a fully automatic factory, much of what is said could be applied to manufacturing firms as they exist today.

THE OFFICE OF THE FUTURE

In the fully automatic factory of the future, the information generated by the production processes will be automatically gathered directly from these processes. It will be possible for the central computing mechanism to keep a continuous running record of each part that is passing through the plant. If the individual machine control circuits are connected with the information-handling circuits of the office, an up-to-the-second, running record can be kept of the position of each part in the production process and of the number of acceptances and rejections at each stage in manufacturing. If a machine breaks down, the production of the other machines and the flow of product units between machines automatically will be adjusted by the computer to compensate for the disruption.

By continuous monitoring of the inspection devices, machine failures can be reduced. The computer will take cognizance of the fact that trouble has begun and either correct it or inform the maintenance crew of the location of the trouble—all of this before the machine actually breaks down, or before many parts are spoiled, or before machining time is wasted on parts that will subsequently be rejected.

Production scheduling and production controls will be entirely automatic. Starting with the product output requirements of the plant, which are introduced into the computer machinery on the basis of management decisions, the computer automatically will schedule the entire production processes of the plant and will take into account disruptions that occur in actual operation. By programming the machine with information now contained on parts lists, specification sheets, operation sheets, and other forms used in production control, it is possible for the computer to determine the optimum runs for each production line, taking into account such variables as the cycle time of individual machines, machine availability, special requirements of certain components, and the probability that delays will occur in specific areas.

Under the most progressive present systems of production control, parts lists and specification sheets are in the form of punched cards. These cards are processed by office machinery to determine the materials requirements for production. With computer circuits, not only will parts lists be compiled, but this information will also be used to determine the optimum scheduling of all production machines. Such automatic scheduling, it should be borne in mind, is done in terms of criteria which are built into the computer's programming. The enormously complicated problem of programming a computer for tasks of this sort soon becomes evident.

A MACHINE YOU CAN PURCHASE TODAY

Lest all of this seem too unreal, let us look at a small computer, costing $15,000, which is currently available to industry for determining production schedules and operators' pay allowances. This device is manufactured by the Potter Instrument Company. Although now being used to schedule production for manually operated, semi-automatic machines, it is performing a task of the same basic nature as will be performed by the production planning portion of the central computing mechanism of the future automatic factory. The following is the announcement made by the Potter Instrument Company when this machine was put on the market.

A new electronic machine now imitates the actions of factory machines and their human operators, at a tremendously accelerated rate, for rapidly determining production schedules and operators' allowances.

Created by the Potter Instrument Company, the world's first commercial "time study computer" simulates machine and operator actions through electronic tubes and can "perform" a day's work in seconds. *Used to study frequently-changing combinations of automatic machine cycles, it calculates the production on each machine of a group* and the waiting time during which the operator is not loading any machine. So that he will not be penalized for an inefficient cycle, the operator is paid an allowance for this time under some modern incentive system.

In a factory employing a number of automatic machines which change their loading and running time cycles frequently as individual job lots are completed, the computation of this allowance has required a large staff of statisticians. This tedious work is now eliminated by the electronic device which completes pay roll or production studies much faster and with unbiased accuracy. Longer work periods can now be studied in a few seconds. The computer is operated by one girl who sets the dials from work slips. No more skill is required than for an adding machine.

As many as ten different machines are "operated" at one time. Each individual loading and running time which has been determined by usual time study methods is set into the computer, in decimal minutes. A press of a button, and the computer simulates operator and machines for any preset period, at a speed of 1,000 elements per second. Where times are carried to thousandths of a minute, the computer simulates an hour's work in less than sixty seconds. If times in decimal hundredths are inserted, ten times this speed is achieved without sacrifice of accuracy, since electronic computation by the "digital" system is absolutely accurate to any number of decimal places required. Choice of operator's loading sequence is provided, and an automatic allowance for a higher than standard work rate where applicable.

Built in three panels, each six feet high and two feet wide, the controls can be reached from a seated position when arranged in a semi-circle. The time study machine operates on a regular light circuit, costs approximately $15,000, takes three months to build, pays for itself in six months.

Any machines, such as milling, drilling, tapping, grinding or processing machines, having an automatic cycle and requiring periodic attention for loading, adjustment or inspection, lend themselves to analysis by the time study computer. Since an attendant can service only one machine at a time, expensive machines may stand idle and production suffer because of inefficient cycles where more than one machine requires attention simultaneously. The possibility of exploring these combinations in advance through the speed of the new electronic computer represents a great leap forward in the production engineering and time study fields.

The enormous complexity of the production scheduling process is too easily overlooked by one not having close contact with modern production processes. The foregoing description of the Potter computer implies something of the work necessary merely to establish the most basic elements of effective production planning—the cycle time necessary to machine specific parts. A completed product unit may contain hundreds or thousands of individual parts. Since these parts must be either purchased or made by the company manufacturing the end product and must arrive at a specific assembly point at precisely the right time, the enormous complexity of modern production scheduling becomes further apparent.

If the reader has toured an automobile assembly plant and marveled at the appearance of red seat cushions at precisely the moment the red body passed the sub-assembly line, and at the arrival of the ton-and-a-half truck body in time for the truck chassis, just missing a sedan chassis, he has had a glimpse of the problems of production planning. Even a manufacturing plant employing a few thousand may have several hundred people planning production schedules. Setting up a production control system, although an enormously complicated procedure, is relatively simple compared with the process of allowing for the variations and exceptions to normal procedure which occur every day. It is these exceptions and special orders that give nightmares to plant managers. A substantial saving would be available through automatic production scheduling equipment, even after allowing for the great difficulty of programming the computing mechanism.

AUTOMATIC ORDERING AND INVENTORY CONTROL

With a properly designed automatic production control system, orders for raw materials and parts can also be automatically prepared. The same information used in determining the optimum runs of specific product units is employed in determining the materials needed by the plant for completion of these runs. With automatic information-handling equipment it is possible for the machine simultaneously to produce lists of required materials as production runs are determined; with proper programming of the machine, these materials requirements can be automatically typed on requisition forms.

Under conditions of steady supply from a small number of vendors, orders can be placed automatically with the vendor by teletype, controlled always by such factors (built into the computer circuits) as present inventory and future requirements. (In an age of many fully automatic factories, it would be possible for the supply requirements of the plant to be directly and automatically integrated into the production planning of the vendor plant.)

Taking physical inventory on an automatic basis is less feasible. The circuits can keep cumulative inventory records by recording how much has been put in and how much has been taken out by specific production processes. However, they will not report discrepancies caused by pilferage and other unrecorded changes in inventory. Various devices such as television screening tubes and the radioactive tagging of individual pieces have been proposed, but as yet there is no practical solution to the problem of putting physical inventory-taking on an automatic basis. However, automatic manipulation of inventory information gathered by human means (by physical inventory-taking) and of information automatically gathered by the production line is both feasible and probable.

A frequently expressed fear is that with fully automatic control of a plant, exceptions to routine (and exceptions are certainly the rule in even the best operated plants) will completely negate the value of the control equipment by making necessary a duplicate set of paper records. This need not be the case. Legal requirements can be met by filming the original documents. Auditors will be able to make use of the computer circuits as they now make use of punched card systems. With proper programming it is entirely possible that all of the information stored in the machine as a result of the production processes can be available to management on a nonrecurring basis.

For example, if it is necessary to know the state of production of a specific part, or of a certain order, it is possible for this requirement to be introduced into the information-handling machinery. A search of the tape can be made automatically, and the part number and the present position of the part can be typed out on an automatic typewriter. The ability to answer such unique requirements does mean considerable complexity in programming; yet it is possible and will be necessary if a computer system is to become productive.

WHAT HAPPENS WHEN A TUBE FAILS?

With virtually all information relating to the operation of the firm contained in the computer circuit, there arises the question of what happens when a tube fails. With no written record of this information, will the failure of a tube, or of the electric current, result in an impossible mixup?

Failure of tubes and of power occurs regularly in normal operations and raises problems that are serious but not insurmountable. Tube or power failure need not disrupt operations or, as some fear, erase all record of them.

As transistors replace vacuum tubes, the problem of failure will be diminished. Even with tubes it is possible to follow the preventive maintenance procedure currently used in the large scientific computers. In short, failures on the part of the machine can be provided for in the design of the machine itself.

Human errors, occurring in the introduction of information into the computer circuits, are at least as pressing as the problem of tube failure. Yet here too the system can be designed to minimize the consequences. When information is transferred from an order form into the computer by means of a typewriter that automatically codes a metallic tape while it types, it is possible to recheck the typewritten data against the order form. Or the information can be coded twice on different but coinciding channels of the same tape and the contents of the channels automatically compared. This is essentially what is done now when two cards are punched from the same data.

INTRODUCING INFORMATION INTO THE COMPUTER

Computing circuits are useful partly because they operate at a high speed; hence, their effectiveness is severely reduced if information must be introduced into the circuits at the speed of humans and if information cannot be taken from the circuits except by slow typing and printing. As has been mentioned, one way of getting around this difficulty is to use intermediary mechanisms; that is, to put information onto a magnetic tape at the speed of a typist and to play this tape into the computing circuits at a much higher speed and to take information out of the computer by playing the tape back slowly to an accounting machine or automatic typewriter. In this way, the operation of the computer itself is not hampered.

The real problem is to eliminate the need for the typist. As more information is channeled directly from the sensing devices into the information circuits, the need for human clerical work will diminish. Information arising outside of production processes, however, will have to be introduced into the circuits of the computer. Ideally, if communications within the firm and between firms were in the form of punched cards or on reels of metallic tape coded and intelligible to the computing mechanism, they could be introduced into the computer simply by placing the reel of tape or the punched cards into the company's computing equipment. This, of course, would require a large number of firms to operate the same type of equipment, which might not be practical. Also, a large portion of the correspondence of a business concern is with individuals and firms not using computing equipment. In addition, it is important that much outside information be in a form legible to management. These facts being true, a wholly desirable development in this area would be a mechanism that could read a letter or invoice or parts list and automatically code the information into the computing machine.

As impractical as such a reading machine might seem, scientists are working on the perfection of such a device. An experimental model of an Analyzing Reader invented by David H. Shepard has been developed by Intelligent Machines Research Corporation of Falls Church, Virginia. The commercial development of this device has been taken over by IBM. The Analyzing Reader scans printed or written material and translates it into a code usable by a computer. The material to be read is fed into the device automatically and read, character by character. Each character is scanned, analyzed, and its identity determined. The reading of a character by the device may initiate any number of actions. For example, a key can be actuated, a portion of programming tape may be set in motion, or the character may simply tell the Analyzing Reader what to do next, such as to proceed to another sheet of paper. The experimental model of this reader operates at an extremely slow speed (one character per second with an error rate of approximately four per cent). It is expected that the production model will operate at speeds in excess of 100 characters per second.

The difficulties of practical operation of such a device are enormous. Information must be printed or typed with a standard type face, or else a new wiring pattern must be plugged in. Howard Aiken of the Harvard Computation Laboratory has suggested that memory circuits be built to handle all known type faces. This could undoubtedly be done, but it is questionable whether it would be practical even in limited form. Nevertheless, various adaptations of such a sensing device will probably be used for introducing information into the computer in a number of different ways.

MORE USEFUL INFORMATION FOR MANAGEMENT

If the output of the computer circuits is in a form suitable for use in making management decisions, the computer is being used far more effectively than if it is employed to tabulate data that must be further manipulated by clerks before becoming useful to management. At the present time much lower and middle management time is devoted to processing information and drawing from raw data the significant facts necessary for making top management decisions.

When a businessman is told that the new technology will provide him with much more information about the operation of his firm, he often shudders, because his desk and briefcase are already crammed with information he does not have the time to digest.

But, if used properly, the computer can do more than tabulate and print increasingly detailed versions of present reports. Used with proper understanding and insight it can give management considerable relief from the perpetual dilemma of wanting more information, yet not having time to use effectively what is already available. The computer can answer the many "What would happen if . . . ?" questions that cannot now be answered. Management can create projected operating data according to various hypotheses. By analyzing these hypothetical reports of costs, production, and profits, under various sets of operation conditions, much can be learned about the wisdom of alternative courses of action.

The rapidity with which even the "slow-speed" digital computers function means that operating statements, analyses of these data, and other special reports can be scheduled and prepared daily with little effort. If need be, the computer system can be run throughout the night by a few operators and provide detailed, literally up-to-the-minute reports each morning. The results of using small IBM and Remington Rand computers for the compilation of financial statements and pay rolls already indicate that such speedy preparation is entirely possible. Fully integrated, automatic information-handling systems should substantially eliminate the present inability to obtain crucial operating information in time to make needed decisions on an intelligent basis.

OPERATIONS RESEARCH

The computer's rapid processing of large quantities of information will make it possible for business to use many of the techniques of *operations research* for both continual and nonrecurring analyses of operations. The difficulties of structuring problems mathematically, securing adequate data, and processing these data have until now considerably hampered the formal application of operations research techniques to business situations, although the techniques have been used from time to time under different names.

Operations research is basically a method of providing quantitative data for the making of management decisions. The needs of the military during World War II for quantitative determination of the effectiveness of alternative deployments of military equipment required considerable time and effort. As a result, a formal methodology was developed, not limited to analysis of past actions but including quantitative prognostication and suggestions for profitable variations in action. The military problems analyzed by operations research teams (or *evaluation* groups) ranged from such simple situations as the formation of mess lines to the establishment of bombing patterns, optimum rate of fire for guns, and determination of the most effective convoy defense arrangements.

Ever since World War II there has been talk of applying operations research to business and other nonmilitary problems. Philip M. Morse, former Director of Research of the Weapons System Evaluation Group, and George E. Kimball, former Deputy Director of the Navy's Operations Evaluation Group of the Office of the Secretary of Defense, wrote a book, *Methods of Operations Research,** presenting the approach and the methodology of operations research and outlining the way in which it can be used by both government and business by creating operations research teams as staff arms of line executives.

Although about a hundred companies are currently using operations research techniques as an aid to making management decisions, there have been many ob-

* John Wiley & Sons, Inc., New York, 1951.

stacles to their effective industrial use. For example, even though many of the engineers and scientists employed by industry are perfectly capable of handling the mathematics of differentials and prediction, management on the whole is not. The advantages of a mathematical analysis of business problems are often not perceived. If it is suggested to management, the arguments are raised that the firm has no one capable of doing such an analysis, that there are insufficient data, or that overly long procedures and too much paper work are needed to arrive at what should be common-sense decisions. Yet the common-sense courses of action often have a surprising way of not turning out, upon analysis, to be the best. But the arguments about involved solutions to even simple problems have much validity. It is here that the use of computers will make the application of operations research techniques far more practical for many businesses than is the case today.

Much of the data necessary for the effective use of operations research techniques are now available to businesses in the form of cost accounting, sales analysis, market research, and time and motion study figures. High-speed, automatic, information-handling equipment makes possible the rapid and automatic accumulation, comparison, and mathematical manipulation of this information. It eliminates one of the biggest handicaps to the present use of these techniques—excessive paper-work demands on already heavily burdened accounting and other office staffs. In addition, the computer permits *experiment,* which has until now been denied to operations researchers working on business problems.

Who in a business organization is capable of structuring problems in the required mathematical terms?

For some problems, particularly those of a nonrecurring nature, it is probably desirable to use a consultant. But day-to-day analysis of operations need not raise serious staffing problems. One or two good men, backed by the present staffs and an analytical approach, should be sufficient to translate the questions executives would like to ask into terms meaningful to the computer. Capable men will be needed, men trained in both computer and operations research techniques. At present such men are scarce, and they command high salaries, but increasing numbers of young engineers and mathematicians are realizing the opportunities in the field. If businessmen make effective use of the capacities of the engineers they now employ, establishment of an effective operations research group should not be difficult.

Operations research, however, should not be considered a cure-all for business problems. In the excitement of their early development, time and motion study, cost accounting, statistics, and human relations were all hailed as final solutions. They all have helped. Yet none is sufficient alone. Likewise, operations research should be viewed for what it is. It is no solution to all the problems of management. It can be of significant help to management by providing quantitative data helpful in making management decisions. Furthermore, automatic information-handling equipment will permit much more widespread use of the operations research techniques.

Before computers can be used effectively for operations research or any other business functions, considerable analysis must be made of the nature and role of information in modern business. If control circuits are to use the information that is generated by the production processes of a firm, not only for control of these processes but in accounting, purchasing, financing, and other ways, something more is needed than appraisal of the efficiency of individual procedures and their translation into computer terms. From the standpoint of effective use of computer circuits, it is no longer fruitful to sort records and forms in terms of department of origin such as finance, accounting, purchasing, inventory, quality control, sales, and personnel. Rather, what is needed is analysis in terms of (a) whether the information originates outside of the firm or is generated by internal operations and (b) the functions common to the processing and use of different kinds of information —similar to *unit operations* analysis. Analysis of this sort is not easy and involves many conceptual problems.

6

What Will Automation Mean to Business?

LEAVING DISCUSSION of the more general social and economic effects of automation for the next chapter, what will be its specific impact on business? Can small businesses take advantage of automation, or do the research and development costs limit the new technology to the industrial giants? How will the increasing importance of engineers change their status within the firm? Although there may be a little labor displacement in total, how does a businessman go about installing automatic equipment in a plant where labor *will* be displaced? What chance is there to automatize when a firm's labor is organized? These and other related questions come to mind when a businessman is told of the great industrial potential of the new technology. Clearly there can be very few general answers to these questions. The specific circumstances of each situation, the economic and political environment, the people concerned, all these will determine the effects of automation in any specific case. Yet, the fact that automation is not new but rather a continuation of a longer trend means that we can learn much by observing industry today and its course during the last generation.

The answers to the foregoing questions and to many others which the businessman may raise can largely be found by observation of the changes that are occurring around us, and that have been occurring during the last several decades. In terms of economic effects, automation is simply continued mechanization. We have already had much experience with the process of mechanization, and it is in this wider context—as a phase in the longer continuum of mechanization—that the business and economic effects of automation must be judged.

The technology of computers and the rapid communications and information processing equipment which this technology makes possible will permit far more extensive analysis of our national economy and business methods, through up-to-the-minute market surveys and extensive research on distribution data and transportation patterns. In such matters as industrial concentration, capital requirements, industrial research, and the utilization of our natural resources, automation will continue trends that are already under way rather than establish new patterns. The growing ratio of capital equipment per worker, the increase in the proportion of indirect to direct labor, and the rise in quality of product (which has led many firms to institute automatic controls even with little or no cost saving), these are all observable trends that automation will accelerate.

LABOR RESISTANCE

Nevertheless, specific problems in regard to automation will trouble the businessman. Perhaps the most pressing is that of labor resistance.

Introduced in a new plant in a new community, resistance will be far less than in an old plant with consequent loss of jobs to many workers in the community. So, too, as products change, there is less resistance to the building of a new plant for the production of a new product on an entirely automatic basis than there would be to convert production of an old product to an automatic basis. Even in those industries having the worst labor relations, therefore, a long-run possibility for automation exists.

As long as the economy remains as dynamic as it has been—and there is considerable expectation that this quality will be with us for a long time to come—there will be less resistance to mechanization than there is in some countries. Organized labor has changed considerably during the last generation in its attitude toward mechanization and continued technological development. Union leaders have seen the need for development and the benefits that it brings to labor. In some industries, of course, organized labor has resisted all technological changes by feather-bedding and make-work tactics. On the other hand, John L. Lewis has criticized the British coal miners for opposing new technological methods, while citing the wage increases which the technological advances of our own coal industry have made possible. It is to be hoped that his attitude will be adopted by all of organized labor.

When automation makes obsolete a specific human skill, it imposes hardship upon the individual laborer who is displaced. In these cases automation will be vehemently opposed.

Even when companies are willing to retain and re-employ workers for other tasks within the same firm, or when other job opportunities exist, there may be considerable resistance to the introduction of new machines in plants where poor labor conditions traditionally prevail. There is obviously no over-all panacea for such a condition. The introduction of automatic machinery cannot be considered a unique occurrence in industrial relations. Rather, the relationship between management and labor is a long-term affair, and depends for its success upon the handling of a multitude of specific day-to-day problems. There is a growing body of evidence that, as might be expected, management-labor relations depend very much more upon the manner in which these day-to-day problems are handled, and the degree of confidence that grows between labor and management than upon any "statement of policy." When these relations have been poor because of lack of understanding or mishandling of day-to-day situations, there is no easy, sure way of introducing automatic machinery into a plant without avoiding labor trouble. On the other hand, when good relations exist because of good day-to-day handling of management-labor relations, automation is far easier than many would suppose.

The Whistle at Eaton Falls, a film written by Sterling Livingston and Paul Ignatius, for Louis de Rochemont, provides an excellent example of the problems which face both management and labor with the introduction of automatic machines. As much as automation may be a new and distinct phase of mechanization, its effect upon workers and upon the economy as a whole is nevertheless distinctly part of the mechanization process. This being the case, there is a great body of experience and theoretical writing to which we can turn for guidance both as to what problems we can expect and as to the manner in which these problems may be solved.

During recent years there has been a growing realization on the part of businessmen that some of the responsibility for retraining and otherwise restoring the earning power of displaced workers lies with management, as well as with the worker. During the last decade many companies trained displaced workers for other tasks. But these were boom times. As long as labor remains scarce this may go on, but we have no assurance that general unemployment will not recur. Although many firms have planned for the introduction of new automatic machinery by giving vocational training and guidance to the workers who would be displaced, business cannot assume the full responsibility for moving such workers to other labor areas and tiding them over financially during their readjustment. This responsibility must be shared by both labor and management— by the nation as a whole—for manpower is part of our national resources.

To some extent training for new jobs can be provided by the educational system of the country. As is brought out in Chapter 7, children are today being trained for jobs that will not exist in the world in which they must earn their living. Likewise, they are not being trained, in many cases, in the basic abilities which are necessary for easy adjustment to the rapidly changing technological conditions through which they will live. If we are to attack the roots of the problem of human resistance to technological change, considerable attention must be given to the upgrading of our educational system so that young people entering the labor force will possess the abilities necessary to adjust themselves to industrial change.

THE ROLE OF ENGINEERS

The present acute shortage of engineers, despite the fact that they have increased from forty thousand to four hundred thousand in the past ten years, makes

it quite apparent that the technical manpower requirements of industry have not been met. Also, as is brought
out in Norbert Wiener's *The Human Use of Human
Beings,* the quality of the engineers currently entering
industry is considerably below the level needed for
effective and rapid technological growth. There is a
great dearth of design engineers capable of the advanced
mathematics necessary in much of the automatic control
system work. It is on the mathematical description and
analysis of highly complex processes and control systems
that the answer to the successful development of the
control systems depends. Automation will further increase the demand for good engineers. In addition, it
will continue and accelerate the placement of engineers
in management positions. This in itself places a further
burden upon our educational system. Increased social
understanding is necessary on the part of our engineers.
They must understand the economic and business context in which they work far more thoroughly than is
possible through "survey courses" in the social sciences
and humanities.

It may be possible to alleviate the problem by attacking it in reverse—by giving management personnel the
technical training, or at least as much technical training as is necessary to understand the methodology and
the basic principles of the engineer. In that way, we
may avoid siphoning off into management the engineers
whose major talents, in design for example, are required
at the technological end of business.

The question of income is of major importance; it
is the opportunity for higher income that lures many
engineers into the management area, where basic design
abilities are not fully utilized. A top-notch Ph.D. can
be hired for $12,000. He is not unaware that he is often

working with management personnel earning two or three times more.

Good design engineers who enjoy their work often realize they are sacrificing potentially higher incomes by limiting themselves to design engineering. This raises the question whether engineers should be compensated on a basis more commensurate with their contribution. The increasing dearth of young engineers has raised the starting salaries enormously, but often at the expense of later salary increases. Salaries of the older engineers, although they have been increased, have not risen proportionately.

It should be clear that, as the direct labor force decreases in number and as premiums are placed upon trained workers and engineers, pressure will be brought by engineering groups and societies for a larger slice of the pie for which they are providing many of the vital ingredients. Although most engineers prefer to remain unattached to labor unions—while reaping the benefits of union-won increases—they are going to require a clearer definition of their status in the organization. Management would do well to give serious attention to the question before it becomes acute.

WHAT FUTURE FOR SMALL BUSINESS?

The demand for engineers and engineering ability raises the question whether any but the hundred or so largest industrial concerns—which have the greatest proportion of scientific personnel—can effectively take advantage of automation. To many small businessmen the new technology may seem likely to work to their own disadvantage. To some extent these fears are justified. In certain fields enormous concentrations of engineering ability and large development expenditures are necessary. It has been apparent for some years that, for the most part, the days of the lone inventor are over. This is true, for example, of the major strides in the chemical industry. But a surprisingly large number of new products is marketed each year by small concerns. This is especially true in the field of electronics—where, encouragingly enough, considerable research and development are required. In fact, the prevailing philosophy in the new product departments of some of the largest concerns is that development attention should be given only to those products not easily produced by small competitors—for licensing of patents has been deemed by many large concerns the wisest way to minimize the danger of anti-trust prosecution.

As flexible, standardized, automatic materials-handling equipment is developed, companies will manufacture such equipment for sale, or lease, in the same way that controls themselves are manufactured by companies such as Minneapolis-Honeywell. Each firm will not have to do its own basic research and development on the application of the new technology.

The development of control mechanisms is considerably ahead of the development of automatic equipment. Using the former as a guide to the latter, we see that very few individual firms have to design their own instruments or control devices. The number of companies making instruments has increased from 684 in 1935 to 1,363 in 1950. As the technology is applied to materials-handling equipment and to standardized fabricating machinery, these products too will be manufactured on a standard basis by concerns wishing to sell such equipment. Research and development requirements, therefore, are not so high for the firms using automatic equipment as they may appear at first.

There is, of course, demand for effective analysis of a firm's present operating procedures and of the changes necessary to make them suitable for installation of automatic equipment. Such work usually requires engineers. Then, too, the machinery itself must often be adapted to the requirements of the firm, although much of this can be done by the companies manufacturing the automatic equipment. Reliable consulting firms can do much of the analysis of present operations and the adjustment of these operations to automatic control.

The basic research on controls, and to some extent on the industrial use of these controls, is being paid for primarily by military and other government funds. If expenditures of this sort should suddenly stop—a very unlikely possibility for at least the next several years— such research would have to be carried on by some other means, but it is clear that very few industrial concerns could afford the immense expenditures for basic industrial research of this type. Establishments such as the Bell Telephone Laboratories have provided some of the most important research, but there are few enterprises of this type in the country. A growing number of industrial research institutions are, however, supported by various companies that provide funds for specific research projects. It is upon such facilities and upon university laboratories that reliance must be placed in the future if effective industrial use is to be made of the new technology.

INDUSTRIAL CONCENTRATION

Even though much of the research in connection with the development of automatic equipment is carried on by concerns who intend to manufacture that equipment and even though common laboratory facilities exist for the use of several industrial firms, it is necessary to ask: Will the capital requirements of automation be so great that the small manufacturer will not be able to afford an automatic plant? And if the small manufacturer cannot afford to use the new technology, will he be squeezed out of the economic picture by large competitors who can? In short, will automation mean only that the rich get richer?

Again we can find the answer by looking about us, for as has been repeatedly emphasized, the economic effects of automation are not new, but rather are a part of the longer continuum of mechanization. Grim pictures of the consequences of industrial concentration have been painted for years by economic pessimists. However, as has been pointed out most recently by John Kenneth Galbraith,* the economy is still operating rather well, and skeptics who have survived thirty years of education in this country still manage to drum up enough intellectual freedom to criticize our economic system.

Automation itself will not produce any startling increases in the degree of industrial concentration. To be sure, the capital costs of an automatic plant will be high—just as are the costs of all plant and equipment today. Certain industries, as is the case today, will be barred to all but those having immense capital resources. But opposed to this there is much evidence that decentralization in ownership, as well as in physical plant, may play an increasingly important role in shaping our economy during the next generations and that industrial concentration will no longer be the bogey it once seemed. This will be true partly because we are be-

* Galbraith, John Kenneth, *American Capitalism: The Concept of Countervailing Power*, Houghton Mifflin Company, Boston, 1952.

ginning to realize that small industry *can* survive along-side industrial giants and partly because there are positive forces at work in the economy to further the role of small business.

The growing use of electricity as a prime power source and the introduction of light new materials are two important factors working toward decentralization rather than centralization of industry. No longer is it necessary to have a very large operation in order to justify the use of an efficient power source. Certain fabricating operations can be carried on in very small plants by the use of electric power as the prime motive force. The fact that automation requires less direct labor means that automatic plants need not be placed in concentrated labor markets but may be situated in far less densely concentrated areas. As flexible automatic equipment is developed and appears on the market, it will be possible for smaller concerns to operate on an automatic basis and to compete very effectively in price and cost with larger concerns in a great many fields. New materials, such as the plastics and light metals, also allow smaller concerns to grow and operate efficiently in specific areas of the economy. The increase in the number of small firms since the war, especially firms manufacturing new products, is an interesting commentary on capital requirements of manufacturing concerns in a new and rapidly expanding era of technology.

Perhaps the most hopeful example of small firms operating aggressively and efficiently in the area of new advanced technology is the group of companies sponsored by the American Research and Development Corporation. This corporation, headed by General Georges F. Doriot, was founded after World War II by Senator Ralph Flanders and a group of farsighted businessmen for the purpose of channeling capital into new ventures —ventures principally based upon technological innovations. The backing by American Research of a group of sixteen corporations, all small, is one of the most hopeful signs in the American economy today. Not only is American Research showing that small companies can still play an active, aggressive role in a highly technical area of the economy and live side by side with well-managed, capable, large firms, but it also provides a thoroughly workable example of the way in which the equity capital shortage of recent years can be overcome.

Other measures are necessary to foster economic conditions favorable both to the continued existence of small enterprise alongside large concerns and to a rate of technological advance in industry commensurate with basic scientific developments. For example, one problem demanding early attention is the tax treatment of corporate depreciation. Injustice is being done to the future generations of this country by the present tax treatment of depreciation. Current depreciation allowances hinder the effective and rapid technological growth of our corporations. This unenlightened treatment actually has imposed a penalty on maintaining industrial plants in effective, modern, operating condition.

In the age of automation we shall reap the benefits of efforts such as those made by American Research and Development and other equity capital companies. Already, new equipment based on the technological developments of the last ten years is being produced by small companies that have been backed by groups of these farsighted men.

7

Some Social and Economic
Effects of Automation

IT SHOULD be clear from the very nature of the
new machines that they will bring important social and
economic changes. These changes require careful study,
but a reasonable analysis of them is impossible before
some attempt is made to answer two very important
questions:

> *How far will automation progress?*
> *How fast will automation take place?*

This chapter does not pretend to offer a definitive
or even a detailed answer to either of these questions,
but it does dispute the extravagant claims of certain
writers. Briefly stated, my answers to the foregoing questions are: (1) Automation will not progress as far as
the proponents of a completely automatic society have
predicted, and (2) the changes will not occur as quickly
as most forecasts have led us to believe. There is enough
evidence in the nature of the technology, in the changes
that have occurred in industry, and in the industrial

problems themselves to indicate, at least in broad outline, the limits to automation and the speed with which we can expect the changes of automation to occur.

As was pointed out in Chapter 2, the difference between the automatic controls made possible by the technological developments of the last ten years and the automatic controls of the preceding hundred and fifty years is that the new technology makes possible the application of self-correcting, feed-back control on a widespread basis, whereas previously such control was possible only in limited uses, such as on a ship's steering engines. This means that the mechanization of certain industries can now advance to a higher level. At the present level, human physical labor is replaced by machine power. At the higher level, the monitoring and control tasks now humanly performed will be done by machines. In addition, automation will mean that certain tasks can be mechanized that previously would have required expensive and elaborate control devices. Mechanization will also take a different form in some cases because of the new technology. But we have explored the possibilities. The important question now is: *What are the limiting factors of automation?*

The most important single factor in determining the extent to which automation will eventually progress is the cost of *taping*. In Chapter 2 we saw the enormous task of putting the problem into terms meaningful to the machine and of building in the criteria necessary for making all decisions that must be made. Once properly taped, the machine can perform a similar task or a varying task again and again. If the machine is to perform this task a great number of times, it is likely to be economical to tape the machine for automatic operation. If the machine is to perform a task a limited number of times, or only once, it may not be economical to tape the machine. The question of what constitutes a long or short run varies considerably with the type of control device, the type of tape, and the product.

The use of General Electric's Record Playback Control (Chapter 4), makes it possible to tape a machine by first performing the action manually. If such control is proved feasible over a wide area, automation will be possible for many job-shop operations where a small number of units must be handled. The most probable course is that controls similar to G.E.'s Record Playback Control will be usable in some job-shop operations but not, for instance, at the corner garage.

Similarly, the economy of connecting machines by automatic transfer devices depends upon the length of the run of a particular product. With the development of flexible materials-handling devices and of reasonably flexible automatic machine tools, the cost of converting a plant from one product to another will be decreased. It will no longer be necessary to have a very long run of a uniform product in order to justify an automatic factory. The length of run required to justify the investment in equipment, including cost of conversion from the previous product, will be decreased. The extent of the decrease will depend upon the degree of flexibility built into the machines. The most reasonable expectation is that medium to long runs of similar products will be susceptible to automatic or almost automatic production.

The new technology will allow the development of new kinds of individual automatic and semi-automatic machines, but the product will still have to be moved from one machine to another. It is unlikely that these machine tools can be connected to one another by materials-handling devices unless a reasonably long run of product is possible.

The extent to which capital equipment will replace labor depends on many factors; in a large degree it depends on the relative costs of the two elements. These costs will vary with the progress of automation.

The foregoing account has been written primarily in terms of our own national economy. In the United States today shortage of labor is one of the principal factors inducing rapid industrial automation. Labor has nearly always been relatively scarce in our economy. It is especially scarce now and is moreover organized to insure itself an increasingly larger share of the gross national product.

How will automation affect employment in manufacturing? While *manufacturing* may seem synonymous with *industry,* manufacturing does not account for even the major portion of our employment. In 1950, 14,884,-000 people, or slightly less than twenty-five per cent of the total labor force (excluding the armed forces), were employed in the manufacturing industries. Automation will be possible in only some of the manufacturing industries, considering these by themselves for the moment, and even in manufacturing industries that are fully automatic, a large number of workers must still be employed.

NO "WORKERLESS FACTORIES"

Automatic factories will not be workerless factories. The examples which exist today show this to be true. The atomic processing plant at Oak Ridge, although operated by a few girls at a control panel, employs many hundreds of maintenance men. Although replacement rather than repair of worn-out or broken equipment is possible in portions of the control system and in portions of certain processing systems, considerable labor is required to disassemble a piece of machinery in order to remove a defective piece, however well designed the machine might be.

One reason for predictions of serious labor displacement through automation has been the assumption that personnel requirements for maintenance would be considerably reduced in automatic plants by the trend to replace rather than repair. This point has been overemphasized. Maintenance personnel in a fully automatic plant with the most extreme policy of part replacement would still be very numerous. It should also be borne in mind that these replaced parts frequently are returned to the manufacturer for disassembly and repair. These operations, although capable of partial automation, require much human labor.

The effect of automation on the information-handling functions of business will probably be more spectacular and far-reaching. Repetitive office work, when in sufficient bulk as in insurance companies, will be put on at least a partially automatic basis. In addition, the machines made possible by the new technology will be used extensively in other areas of human endeavor. For example, mathematicians, economists, military theorists, and meteorologists are already using the computers. Wider uses will become possible when smaller and cheaper computers become available. New devices will be developed for use in medicine, retailing, and many other occupations. When used in these ways, however, mechanization rarely replaces labor, for the process does not become entirely automatic. In most cases, automatic machines will make it possible to render new, more comprehensive and more economical services. In this way there may well be continuation of the present trend toward expansion of the service industries in relation to other industries.

Very few offices will be entirely automatic. Even in the most automatic, it will be necessary to have people to tape the machines. These people must be of a high level of technical competence. Besides, the normal operations of any business will require typists, stenographers, office boys, receptionists, and the like. Under optimum conditions it may be possible to have an office as automatic as that outlined in Chapter 5. Yet, even in such an establishment, it is evident that there will be demand for personnel, technical and nontechnical, semi-skilled and unskilled. Fully automatic offices will be possible in only a limited number of cases. By far the most common situation will be that of using certain types of accounting and information-handling machines as a part of office operation as it is known today. Certain functions, such as filing or statistical analyses on the lower management levels, will be performed by the machines, but much day-to-day work, answering correspondence and the like, will have to be done by humans.

Automation will be limited where human interaction is of primary importance, as in retailing. Machines may be used to perform accounting and some order assembly services in department stores and to perform certain limited vending services at railroad stations and airports and even in retail establishments, but the use of machines in retailing will be greatly limited by the customers. People like to feel, look at, and discuss merchandise they intend to buy. To some extent, packaging and standardization will allow increased use of automatic selling devices. But there are limits beyond which customers do not seem willing to sacrifice individual choice (in expensive cuts of beef) despite the somewhat lower price made possible by standardized packaging. Some supermarkets have compromised by providing both clerk-serviced and self-service counters. It appears likely that future developments in automatic vending will take the same course, rather than tending toward wholly automatic stores.

WHOLESALE ORDER ASSEMBLY

In wholesaling the situation is different. An excellent example is provided by the automatic order assembly system designed for the suspender plant of the Hickok Manufacturing Company at Lyons, New York, by V. H. Laughter, owner of the Code-O-Matic Company, of Memphis, Tennessee. Prior to the installation of the automatic system, wholesale orders for suspenders were assembled at the Lyons plant by an employee who pushed a cart around the stockroom, picking up the items ordered. Because of the hundreds of size-style combinations, the hand assembly of orders represented a considerable labor cost.

Mr. Laughter installed a series of six hundred chutes to hold the merchandise. The chutes are operated by remote-control code machines. The operator pushes keys on the code machine corresponding to the size, style, and quantity desired. The items are then released from each chute by a solenoid-controlled gate. The boxes fall on a fast-moving belt and are carried to a central point for packing and shipping. An entire order can be assembled as rapidly as the clerk can punch the keys (this could be done by prepunched cards if the operation were large enough) . The installation, although custom-built and experimental, has worked very well and has paid for itself in a short time.

AUTOMATION IN RETAILING

Clarence Saunders, who left his mark on retailing by the establishment of the first supermarkets, the Piggly Wiggly stores, experimented several years ago with a fully automatic grocery store. The Keduzal store, as it was called, was built and operated in Memphis, Tennessee. Only items that could be packaged and machine-vended were sold. Each customer picked up a "key," or small piece of paper tape wound on a special handling device, when he entered the shop. All merchandise was displayed in cases, each case holding twenty brands, styles, or types. The customer inserted his key in a machine and pressed a button on the unit displaying the merchandise he wished to buy. The machine punched the tape and simultaneously printed a line of type stating the class of merchandise, quantity, and cost.

After "shopping" in this manner, the customer presented his key to a clerk. The key was inserted in a tabulating machine and the tape was run through a sensing device at high speed. The sensing device operated electrical relays at the ends of merchandise chutes. The chutes dropped the appropriate packages onto a continuously moving belt. The entire order dropped onto a belt in a few seconds and was collected in a basket at the end of the belt by a clerk. The complete package of merchandise was ready for the customer within two minutes of the time he presented the key to the clerk.

The Keduzal store was a financial failure. There are several suggested explanations. Some felt that development costs had been allocated on too short-term a basis.

Others say too many changes were made in attempts to improve the mechanical devices. One cause, however, was quite clear. After the novelty wore off, customers found the selection of merchandise too limited. It was not possible to do all of a family's food shopping in the Keduzal store. In an effort to overcome this drawback, attempts were made to carry merchandise that was not readily packageable. There were distinct mechanical limits to this, however. The cost of providing attended meat counters, for example, might have largely eliminated the savings of the automatic order assembly system.

If automatic totting of orders—one of the major savings of the Keduzal system—were possible without automatic assembly (since the customer's hand assembly of orders is without cost to the retailer, provided reassembly and checking are not necessary), a workable compromise might be developed. Saunders has developed such a system, dependent upon honest recording by the customer. A simple vending and checking machine might also be used, but for present purposes the importance of the Keduzal store experiment is as an indication that, although some automatic vending is possible for low-cost items, it is unlikely that all retailing will ever be put on a fully automatic basis.

In retailing high-cost items such as fine luggage and clothing, automation is severely limited. This is not to say that retail operation will not benefit substantially from automation. For example, Edward Rogal developed a Central Record System in the 1930's that does for a department store much of what the automatic office described in Chapter 5 does for manufacturing concerns. The further development and use of central record systems and central automatic accounting systems of this type are to be expected. But this is a far cry from a fully automatic department store.

MAGNITUDE OF CHANGE

It is difficult to make a definite prediction of the extent to which automation will progress, either in regard to the number of people whose jobs will be affected or the number of firms that will use automatic equipment. It is, however, possible to obtain a very rough estimate of the extent to which automation will directly affect jobs by considering the types of industries which will be affected. Although they will use automatic machines, agriculture, trade, service, construction, mining, and the self-employed and professional fields will certainly not be automatized. In 1949 these fields accounted for over fifty-six per cent of the total labor force, excluding the armed forces.

Agriculture will be changed considerably during the next few decades by the production of raw food material through the growth of algae in controlled tanks or plants. Construction will be affected by the perfection of prefabricated housing and the use of factory-made building units. Mining productivity will be aided by the use of new devices based upon the technology of feed-back. The service industries will use machines for manipulating data in market surveys as well as for such simple operations as washing dishes. Proprietors will have better record control systems for their shops, and professional men will be able to render a higher level of service through the use of new machines.

But none of these groups will be automatized in the sense that humans will no longer be necessary or even in the sense that the need for humans will be radically reduced. As in the past, new developments in industry will lead to growth and expansion of the economy so that there will be more jobs even in those industries where automation is most completely adopted, because these industries in most cases will also expand. There will be cases, of course, in which specific hardships will result from the displacement of individuals who possess certain skills, even though the total number of jobs remains the same or increases. This is discussed more fully later.

In an excellent unpublished paper entitled *Automatism in the American Society,* a physical scientist, Richard L. Meier of the University of Chicago lists the industries he believes ripe for automation. They are (by U. S. Bureau of the Census categories) : bakery products, beverages, confectionery, rayon, knit goods, paperboard containers, printing, chemicals, petroleum refining, glass products, cement, agricultural machinery, miscellaneous machinery, communications, limited-price retailing, and some miscellaneous items. These industries use about eight per cent of the total labor force. As Mr. Meier states: "It is of course unlikely that totals in any of the separate categories could be reduced as much as fifty per cent over a twenty-year span, but it should be remembered that some effects will be noted in other categories than those mentioned." These other categories are, of course, some of the information-handling and accounting functions of businesses which are not generally susceptible to automation, as well as the use of computer techniques and some of the new technology in developing machines for industries that cannot be completely automatized.

Nevertheless, the labor shifts that could be expected from these changes and the time span during which they are likely to occur are no greater (Mr. Meier would say smaller) than the abnormally great population shifts that occurred during the 1940's. This comparison emphasizes the fact that the nature and rate of population shifts due to automation are both of an order of magnitude with which we are historically familiar and with which we are able to cope.

SPEED OF CHANGE

The speed with which the changes of automation occur is of considerable importance. If the changes of the last fifty years had occurred within a span of ten years, they would have had notably different consequences.

Automation eventually will make possible important and substantial changes in the ways we earn our living as well as in the ways we spend our leisure. Yet the social effect of these changes depends in great measure upon the *rate* of change. It is therefore important that we give as much attention to this as to the problem of the degree of change.

Mr. Meier, in *Automatism in the American Society,* says:

The most striking conclusion to be obtained from this fairly comprehensive survey of the technological applications of automatism is that the rate of change effected is not expected to be astonishing. Several reasons might be advanced. The most important of these is the fact that all operations involving non-linear differential equations cannot be evaded, and their solution will come slowly. Also, most of the effort is being directed to the military sphere which, because of secrecy and specialized technique, can affect the civilian sector only at a few points (note the slow rate at which radar techniques, perhaps the most useful of all military communications developments, were taken up by industry). Finally the prospective shortage of technical manpower will put an effective damper upon the rate at which new developments can be made available to the public.

Norbert Wiener probably would not go along with this prognosis. He and others have expressed the opinion that in one generation, or two at most, automation will bring about extensive changes in industry and in our way of life. The author cannot help but feel that overlooking the as yet unsolved industrial and economic problems has resulted in an exaggerated idea of the extent and speed with which automation will proceed. Many management, engineering, and organization problems remain to be solved. Where a huge capital investment is required, as in the steel industry, it is entirely unreasonable to expect that the full effects of automation will be felt during the next ten years.

Of course, technological change is not always a slow process. Only eleven years elapsed between the time Jacquard first exhibited his automatic loom in Paris in 1801 and the time 11,000 Jacquard looms were operating in France.

The changes of the last hundred and fifty years have been very rapid as compared with the changes of the last thousand years, for a rate of change is meaningful, of course, only in relation to another rate of change. So, too, the changes of the last fifty years have been comparatively more rapid. And the most reasonable expectation is that the changes accompanying automation will occur at least at the pace of the last two decades and, in all probability, somewhat more rapidly. The pre-steam-age resistance to industrial progress is no longer so strong. To be sure, there is a natural reluctance to change our ways of doing things. This slows up the processes of industrialization and mechanization. Yet we are becoming more and more conscious of our "mental blocks" and are continually taking steps to overcome them.

Despite the rapid pace at which mechanization has progressed during the last fifty years, it takes many years to bring about changes in the total capital structure of an industrial society. One has only to consider the extent of obsolescence in our own country, the foremost industrial power in the world. After one becomes accustomed to the glitter of the truly magnificent industrial advances made in this country, one is struck by the *age* of our industrial plant. Despite increasing monetary investment in industry, more than one of every five machine tools in our national industrial plant is over twenty years old, and forty-three per cent of all our machine tools are ten years old or older. H. E. MacDonald has kindly made available to the author his detailed analyses of industrial obsolescence, and from these it appears that, if the present replacement trend continues, by 1955 more than three out of four machine tools in this country will be at least ten years old.

Although these figures certainly do not reflect a true picture of the state of obsolescence of our national industrial plant as a whole (how many nylon, plastic, or uranium plants existed twenty years ago?) they emphasize a fact that is very important in considering the speed at which automation will progress. Although technological change progresses rapidly and in some sectors of the economy may cause substantial change in less than a human lifetime, many years are necessary to change the total industrial scene. We have experienced many startling technological innovations in the last two generations, but the rate of change has not been too fast for human adjustment.

Even the most revolutionary and startling technological developments are taken as a matter of course by many of us. This is not to say that these changes do not raise problems. But people do seem to be able to take even the most important and greatest technological changes in stride and to adjust to them. Virginia Woolf has expressed this phenomenon very well in the following passage from her novel *Orlando*.* Orlando, who is a personification of the spirit of England, and has thus lived for several hundred years, suddenly finds herself traveling from her country house to London on a train rather than by the horse-drawn coach she formerly used.

> Orlando had not yet realized the invention of the steam engine, but such was her absorption in the sufferings of the being, who, though not herself, was entirely dependent upon her, that *she saw a railway train for the first time, took her seat in the railway carriage, and had the rug arranged about her knees without giving a thought to* "that stupendous invention, which had," the historians say, "completely changed the face of Europe in the past twenty years" (*as, indeed, happens much more frequently than historians suppose*.)

In historical retrospect, what seems a world-shaking change is often accepted by the people of the day as something new that can be lived with, adjusted to, and perhaps even used with some benefit. Certainly in this day and age we are accustomed to the spectacular and to world-shaking changes. This does not mean that we have so organized our society that problems do not arise. It does mean that the changes themselves do not bring about as complete and drastic a revision in the structure of either industry or society as is often expected.

ROBOTS?

I have tried thus far to build a frame of reference—the probable extent and speed of automation—

* Harcourt, Brace & Company, New York, 1928.

within which to discuss social and economic consequences. However, it is necessary to make clear one further point concerning the supposed human characteristics of the new machines. Writers such as Norbert Wiener, by emphasizing the similarity of automatic control systems and the nervous systems of humans and animals, have made the world of science fiction seem indeed to be upon us, with a race of human-like robots already in the making. No interpretation of the facts could be more perverse—or disturbing.

Robots, machines which look and act like humans, have been the subject of speculation and fantasy for many years. *Frankenstein* (1818), *R. U. R.*—Rossum's Universal Robots (1921), and the film *Der Golem* (1915) are but three classical examples which have fascinated many of us. Currently the subject is enormously popular, and the pseudo-scientific language in which today's stories are told, when coupled with the animal-machine analogy of the Norbert Wiener school, surrounds the whole with an aura of reality. Even so recent and serious an account of the new technology as Edmund C. Berkeley's *Giant Brains* warns, "There seems to be no kind of escape possible. It is necessary to grapple with the problem: How can we be safe against the threat of physical harm from robot machines?" (John Wiley & Sons, N.Y., 1949.)

But let us look at the facts. The solution of mathematical problems, and other feats performed by computers, do resemble the processes of human thought. But the resemblance is too superficial to warrant the conclusion that these machines *think* or are in any essential way *human*.

The problem is largely one of semantics. Our language has not yet developed the words that deal accurately with the new concepts. Even the word *computer* is hopelessly inadequate in describing what the machine is. In the broadest sense that we use the word *think,* it can safely be asserted that we have no machine that thinks. What a computer does is to carry out a logical process that is not at all human. To call it "thinking" would be as incorrect as to say that a typewriter has human characteristics because it can "write" this book.

The accounts that describe the new machines in human terms neglect one very important fact. *Free will,* the essential human quality, is absent from all of these machines. In no way can this quality be attributed to any machine yet developed, nor is there any indication that any such machine *could* be developed.

It is true that there are many similarities between the nervous system, the computing machines, and some of the phenomena of control, but it is felt by many competent psychologists and physiologists that much of what Norbert Wiener has said is of questionable accuracy in that the similarities between nervous systems and electric networks are far more superficial than Wiener and his colleagues admit.

To build a machine that will correct its own errors in accordance with criteria predetermined and built-in by humans is very different from creating a machine that is human. Even if it were technologically possible to build machines which could perform all the work that is presently performed by humans and which had the ability to think—and even possessed free will—there is considerable question, aside from all moral issues, whether we would economically want to produce such machines. From the standpoint of performing industrial tasks, humans are, in most cases, very inefficient. Oil refineries and automobile plants bear very little relation to the structure or the function of a human being. Human robots would thus replace humans in our *present plant*. But what a waste of investment this would be! For what a waste of human resources it is at present to have a human being, capable of all a human can do and feel and express, standing in an assembly line tightening nuts! If we possessed sufficient technological ability to develop human-like robots, what a waste it would be to go about our industrial tasks in the way we perform them today. How much better to build machines which could perform these tasks without having the added ability to play games of chess, to walk, to solve difficult problems and to communicate with others. So, too, it would be a waste of investment to build machines—even though not possessing human qualities—to perform the multitude of functions of which a worker is capable. We want flexibility in our machines, of course. We want multi-purpose machines, too. But the maximum degree of flexibility that could be industrially useful and economical falls very far short of anything even remotely approaching the fully human machine.

Automatic machines will thus not take on human form. Nor will they be all-purpose robots. The new technology permits the development of a group of much more automatic machines than we now possess. But such machines will be related to the machines with which we are familiar in the sense that they will perform certain groups of industrial functions and will vary considerably according to the functions to be performed. Even where present technology remains unchanged, as might be the case in oil cracking, the automatic machines that take over the functions now being performed by humans will certainly not be human any more than the electric eye that opens the door in Penn-

sylvania Station is a *Cyclops,* or the group of Minneapolis-Honeywell regulators monitoring the output of
a continuous flow plant is human.

Some of the scientists who have participated in
the development of the new technology have given considerable thought to its social meaning. Many of them,
while fully realizing the revolutionary aspects of these
developments, have also recognized their place in the
longer continuum of mechanization. Unfortunately
some have not. The more vocal of these have caused
considerable apprehension. The following draft passage
from his forthcoming book, *Technology and Human
Values,* which Eugene Staley has graciously made available to me, indicates the extent to which a man who
has done outstanding work on the subject of technological unemployment feels that the social consequences of
automation have been misrepresented by some of the
physical scientists who have written on the subject:

I wish that some of the physical scientists and engineers who
write about automation and its effects would consult a bit with
their brothers in economics and the other social sciences before
they commit themselves to scare-prophecies. Mr. Norbert Wiener,
mathematician and pioneer researcher in control and communication mechanisms in animals and machines, has contributed note-
worthy and stimulating ideas in his *Cybernetics* and in *The Human Use of Human Beings.* But why in the latter book must he
fall into the trap of the old "lump of labor" fallacy and tacitly assume that there is only so much work to be done in the world and
therefore that a startling increase in efficiency will create mass
unemployment? There are some very able economists in the same
building with him at the Massachusetts Institute of Technology,
and a little journey down the corridor for some internal communication could have saved him from setting down his rather naive
remarks on technological unemployment. This is not to say that

there is no problem of technological unemployment, but that there are a good many factors, now fairly well known, which determine the nature and size of the problem under given circumstances. We can draw on accumulated analysis and experience in the social sciences in this area; it is not necessary to repeat all of the mistakes made by previous generations of workers in these fields.

Automation must be viewed in proper historical perspective as a new chapter in the continuing story of man's organization and mechanization of the forces of nature. It raises new problems. It solves some of the problems, human as well as mechanical, that were raised by earlier phases of mechanization.

We have a considerable body of experience and analysis to guide us in determining the most probable economic and social effects of automation. This fact is frequently forgotten. To be sure, automation does have special characteristics that will uniquely affect the nature of the social and economic changes, but it is not a departure from all that has gone before.

If more attention were given to the past by the prophets of things to come, we would have far less fright and fewer confused predictions. If this chapter can be said to have any one message, it is to urge sound writing and thinking on the effect of technological change on human society and to deplore the predictions of the debasement of the human race.

DEBASEMENT OF THE WORKER?

Perhaps the most vehement objection to industrial use of automatic controls—and indeed to any form of mechanization—is the charge that machines debase the worker; that the craftsman's skills which gave stability and meaning to life are superseded by automatic machines that subjugate the worker and deprive him of the opportunity to make his own creative contribution. This recalls the protests of the Luddites, who saw in the steam age an end to our civilization. In varying form these protests continue today.

But when the condition of labor in today's mechanized plants is contrasted with the condition of labor in the past, the burden of proof is surely thrown upon those who decry modern industrial development. In *Work and Wealth,** J. A. Hobson stated:

> There has never been an age or a country where the great bulk of labour was not toilsome, painful, montonous, and uninteresting, often degrading in its conditions. Bad as things are, when regarded from the standpoint of the human ideal, they are better for the majority of workers in this and in other advanced industrial countries than ever in the past, so far as we can reconstruct and understand the past.

As much as we may sympathize with the ideals that lead to criticism of our industrial civilization as depriving man of an environment in which to develop to his fullest stature, the truth of Hobson's statement cannot be denied. It should serve as a benchmark from which realistic discussion can proceed.

Lest the social critic brush off Hobson as one who uses glittering generalities, Abraham L. Gitlow has set down the advantages of our modern industrial society, chapter and verse, in his excellent article, "An Economic Evaluation of the Gains and Costs of Technological Change": †

> . . . improved American living standards, based on technological change and expanded productivity, have meant: (1) higher real wages; (2) shorter hours and increased leisure; (3) more ex-

* Allen and Unwin, Ltd., London, 1933.
† Industrial Productivity, Industrial Relations Research Association, Madison, Wis., 1951.

tensive schooling; (4) a greater margin of family spending on items other than food, housing, and clothing; (5) a longer life expectancy at birth; (6) better working conditions (factory decentralization, less arduous labor, vacations, sick leaves, etc.); (7) improved nutrition; (8) improved housing (central heating, electricity, household appliances, baths, flush toilets, running water, etc.); (9) better family living (a growth of such noneconomic functions as recreation, travel, reading, plus the trend toward suburbanization); (10) improved status of women (more freedom outside the home, plus the great lightening of the once harsh and burdensome responsibilities of the housewife in food preparation, laundering, cleaning, and making and maintenance of family clothing); (11) the ability of the economy to maintain an increasing proportion of the labor force in scientific and professional pursuits (yielding, in turn, a rich harvest in medical and technical research with ever-widening ability to apply the results of such research); and (12) an improved life for farmers (increased incomes and less arduous labor due to more extensive applications of machinery, and less isolated life due to automobile and radio).

Having made clear my feeling about the over-all benefits we enjoy as a result of technology, I would like to express some misgivings, with the expectation that they will be considered in the light of what already has been said. Although the Luddites and their modern-day counterparts have been proved wrong, there is certainly some truth in the fact that the way we organize production in our modern factories does, in many cases, subordinate the human to the machine.

In the spirit of René Clair's *A Nous la Liberté* (1931), Charlie Chaplin's *Modern Times* (1936), with the exaggeration of artistic license, captured the spirit of the age—to mechanize, systematize, and regiment factory work—and, where the work is done by men, to mechanize and regiment the factory workers. Many of the human problems of today are the result of the attempt to adjust the worker to the machine that paces him and, in a broader way, of the mechanistic concepts of the function of workers in *mass production*. The works of Elton Mayo and Fritz Roethlisberger offer ample proof of the severe human problems that have accompanied the process of industrialization.

It is necessary to look no further than the role of the machine operator in a factory. To a large extent, although differing markedly from industry to industry and from plant to plant, industry subordinates the direct laborer to the machine. Machine operators start and stop the machine; they put work into and remove it from the machine; they watch for breakdowns of the machine. In most cases the operators *tend* the machines, while the machines perform the vital fabricating function. In addition, and particularly on the assembly line, the operators are largely *paced* by the machines. Even when the machines are adjusted to a speed slower than the average worker, the operator *feels* paced by the machine. At best this cannot be considered desirable. That it may be far less than desirable has been brought out recently by Charles H. Walker and Robert H. Guest in *The Man on the Assembly Line.** Their studies reveal that assembly line pacing is responsible for a considerably larger portion of the psychological unrest and discontent in industry than apologists of the system would have us think.

SOLUTION OF SOME HUMAN PROBLEMS

It seems appropriate to ask: *Will automation aggravate this situation?* The answer is *no*. Although automation will assuredly bring its own problems, it provides the answer to the human problem of machine pacing and subordination of the worker to the machine.

* Harvard University Press, Cambridge, Mass., 1952.

Automation makes possible the development of machines to perform the repetitive work of industry. In the factory of the future, control mechanisms will monitor the operation of the fabricating machines. Automatic materials-handling devices will not only move the workpiece from machine to machine but will load the workpiece into the machine and remove the completed piece from the machine. Devices more acute than our own sense organs will warn the control system of trouble and, in most cases, the machine itself will correct trouble that arises in routine operation. What this means is that to a great extent the jobs in which the worker is tied to and paced by the machine will be taken over by other machines. The worker will be released for work permitting development of his inherent human capacities.

Taking the opposite view, it is sometimes argued that workers doing routine jobs are happy. The implication is that, given tasks more difficult and demanding, the worker would become discontent, unhappy, rebellious. Actual cases are cited to prove this. But these are usually instances in which the worker was removed from a psychologically satisfying situation to which he had adjusted himself and was put into a new, unstructured, and insecure situation.

Many people are happy doing the simple, elementary, repetitive tasks of industry. But to say that they are happy is not the same thing as saying that they cannot also be happy or happier performing tasks that use their abilities more fully. There are undoubtedly people of extremely limited capacities who would be frustrated and upset by any work more challenging than the most elementary and repetitive tasks, although this group is probably far smaller than is commonly supposed. But even under the best of conditions there will be all too many jobs available for people of limited capacities. It should be clear that we need not adjust industry to the lowest denominator of capability.

In *The Human Use of Human Beings* * Norbert Wiener writes:

> It is a degradation to a human being to chain him to an oar and use him as a source of power; but it is an almost equal degradation to assign him purely repetitive tasks in a factory, which demand less than a millionth of his brainpower. But it is simpler to organize a factory or galley which uses individual human beings for a trivial fraction of their worth than it is to provide a world in which they can grow to their full stature.

I could not agree more fully. It is easier to scoff at the possibility of workers doing tasks that demand a high level of ability than it is to train and adjust them for such work. The humility that management needs in order to understand why employees become hostile or "ornery" is not taught in any school. Yet it is a quality that, although always essential in administration, will be of even greater importance in the future.

LIMITS OF UPGRADING

I do not suggest that the assembly line worker must be transformed into a design engineer. The work that will require the most manpower will be semi-skilled and highly skilled maintenance and repair. Such work, although on a much higher level, is in most cases fully within the ability of the people who today work at the simple repetitive tasks of the assembly line, provided of course that they are properly trained and motivated. The maintenance and repair jobs require a different set of abilities than are needed for engineering and design. A high level of theoretical comprehension is not

* Houghton Mifflin Company, Boston, 1950.

so important in these tasks as are genuine interest in the work, desire to do good work, and ingenuity. In an odd and entirely unexpected way, automation may bring us back to the human and psychological values of the self-respecting craftsman. Electrical and mechanical repair work, instrument adjustment, and general mechanical tinkering can provide challenges, pleasures, and satisfactions very much like those enjoyed by the swordsmith or cabinetmaker of old.

Maintenance and repair jobs are not *paced* by machines. The work is done *on* machines, but that is psychologically quite different from being paced by an assembly line or by a semi-automatic fabricating machine. Moreover, the extent of personal challenge that maintenance and repair work offers should not be underestimated. Anyone who has worked with men performing maintenance and repair functions can recognize the great self-respect they enjoy and the respect they earn from their fellow workers.

In addition to an absolute increase in future requirements for semi-skilled and highly skilled workers, the ratio of maintenance and repair workers to direct laborers will increase sharply during the next two generations.

The *upgrading* of labor that will accompany automation will not be limited to the acquisition of mechanical skills but will be a rounded process of fuller development of the whole man. To quote again from Staley:

Automation will mean that human labor in the advanced technological societies will be used less and less for the routine, repetitive functions involved in feeding workpieces through machines and assembling them into finished products. Labor will be more and more "up-graded" into the kinds of functions performed by the engineer, the designer, the production planner, the skilled maintenance and repair man, the organizer and manager. Social skills, that is, skills in human relations by which people are enabled to work together, will continue to become more and more important in relation to mere mechanical skills. We will have more and more leisure, and more and more time and necessity for thinking about the broad aims of human living rather than the day-by-day problems of getting enough to keep body and soul together.

THE PROBLEM OF LEISURE

It becomes clear that in providing more satisfying jobs that allow fuller human development, automation will heighten the problem of *leisure*. Are we capable of developing a culture that does not depend upon work to give meaning to our lives? As material needs are more readily satisfied by increased productive power and as less human effort must be expended upon production, the question of leisure and its function in our lives will become more pressing. This will not happen overnight. The urgency of the question will depend upon the rate at which the new industrial technology advances, upon whether we have an all-out war, and upon the extent and duration of the cold war armament program. The greater the proportion of production for military rather than civilian uses, the less pressing will be the problem of leisure. Even at the fastest rate of technological development and with the smallest proportion of national output devoted to war-like uses, the problem of increased leisure is not likely to become acute very soon. The change will be gradual and will permit time for adjustment.

That increased leisure should become a problem is, of course, something of an anomaly. However, a problem it is, and it will be with us to an ever-growing extent. We are thus faced with the necessity of both asking and answering the question: What role should leisure play in our lives? The arguments that have been advanced for decreased working hours and a shorter work week have, in most cases, been that the worker is more effective because of these shortened working hours. This is undoubtedly true, and many of the changes of the last fifty years have done no more than provide adequate rest periods so that the worker is more effective on his job. Even as late as 1923, certain jobs in the steel industry required twelve hours of work a day for seven days a week. A change from this to an eight-hour day and a five-day week is primarily a change which makes for more effective performance on the job.

However, is further shortening of the work week and of the hours worked per day necessary for effective performance on the job? More and more the question is raised: Can we as a people learn to utilize leisure as something more than a respite during which we overcome the effects of work and prepare ourselves for additional work? This is one of the most basic problems of our day. It is one which automation accentuates but has not produced, for the problem has been with us, to an ever-increasing extent, since man first found means for providing the barest material essentials of life.*

A CHANGING POPULATION

Two of the factors affecting the speed and extent of the increase in leisure time are, first, the basic changes in the population pattern of this country and, second,

* Josef Pieper in his book, *Musse und Kult* (translated into English by Alexander Dru, and published by Pantheon Books, Inc., as *Leisure, the Basis of Culture*), presents a succinct philosophical analysis of this most important problem. The report on the Corning Conference, *Creating an Industrial Civilization*, edited by Eugene Staley (published by Harper & Brothers, 1952), provides an even more rounded discussion of the problem, and unlike Pieper includes arguments that work can provide a true basis for a full life.

the demands made upon our economy by our struggle with communist totalitarianism.

During the last fifty years the total population of this country has almost doubled. This rise in population would, of course, have necessitated a doubling of physical production in order merely to keep the standard of living at the nineteen hundred level. However, in order to increase the living standard to the extent that we have done since 1900, the increase in productivity, or output per manhour, has had to be more than double.

Productivity has risen at an annual average rate of about two per cent a year, but that rise has been sporadic. Productivity in some industries has increased at a very high rate. In others it has decreased. Year-to-year variations are very great. Industries which in the long term are increasing in productivity pass through years during which substantial decreases take place. It seems very probable at present that we are undergoing a period of decreasing manufacturing productivity. This will be discussed more fully somewhat later in this chapter. But even if manufacturing productivity were not decreasing, changes within our population, other than the increase in total population, have been such that increases in the productivity of the working force will be necessary within the next few generations in order to maintain or increase our present standard of living. Automation is, of course, a means for increasing productivity and, in fact, the increased productivity obtainable through automation is possibly the single most important economic meaning of automation.

The great increase in population during the last fifty years is but one manifestation of the dynamism of our population. Changes in the composition of the work force have been equally great and are of equal importance. The primary change is the increase in the ratio of dependent older persons to the active work force. Sixty years ago somewhat more than sixty-eight per cent of the men over sixty-five years of age were working. Today forty-two per cent of the men over sixty-five are in the labor force. A number of reasons can be given. The population shift from farms to cities during the last half century has put older people in an environment in which it is difficult to work productively in their old age. The increase in the tempo of work has made it difficult for older workers to obtain jobs. With pension systems an employer is less inclined to hire a man past forty because of his short remaining work span in relation to the pension that must be paid. It is harder to fire an older man than it is a younger man who can readily obtain another job. Older people are sometimes less adaptable. For these reasons, younger workers are at a premium in today's industry, and older people are gradually forced out.

To be sure, older people constitute a labor reserve which, with housewives and teen-agers, can be called upon in event of emergency or war. During World War II all three of these groups added substantially to the work force. On the whole, in the absence of emergency or war, we can expect a continuation of the trend toward a higher proportion of older, nonworking people.

The great population increase resulting from the higher birth rate beginning with World War II may suggest that the addition of younger people will more than compensate for the retirement of older people from the work force. There is considerable doubt, however, that the increased birth rate is permanent. The most capable demographers were, of course, proved wrong by the phenomenon of an increasing birth rate in the 1940's. As late as 1941 and 1942 the most reliable population predictions were for a steady decrease in the birth rate to result in an eventual decrease in total population. Population forecasts are obviously quite risky. There seems to be a considerable body of evidence, however, that the increased birth rate of the 1940's is not likely to continue.

The rising birth rate during the 1940's resulted from an increase in the *number* of families rather than in the *size* of families. Most births were first, second, or third children, with first and second children constituting the greatest number. Births of fourth, fifth, and sixth children, although increasing in absolute numbers, did not increase as a percentage of total births. Many births were postponed from the depression, but the rapid lowering in the marriage age accounted for the greatest birth increase.

A trend toward a lower marriage age is self-limiting and would appear to be already at its lowest probable limit. This means that a major factor in the increased birth rate will soon cease to be felt. A rise in the marriage age would mean a very sharp drop in marriages and potential births.

Although recessions may continue to occur, we are not likely to see a recurrence of the depression of the 1930's. With a high rate of national income, the most probable course seems to be a continuation of the present low marriage age, which in turn means that, although there will be some decrease from the high birth rates of the forties, we are likely to continue increasing the total population at a substantial rate. On the other hand, the channeling of youths into the military service for a two-year period, which seems probable for a number of years, will tend to decrease the work force. The problem is complex.

All in all, the most probable net result of all changes would seem to be an increase in the percentage of older, nonworking population in proportion to the active working force, which again emphasizes the need to raise the productivity of those who are active in the work force if we are merely to hold our own, much less raise our standard of living.

THE COLD WAR

The need to increase not only our own productivity but also that of other nations is apparent from another and even more pressing quarter—the struggle between the free world and communism. Unless there is war, and perhaps even then, it seems improbable that this struggle will be settled conclusively for a long time, perhaps several generations—the time during which automation will produce its greatest changes. It is to our advantage that we do not stand alone in this struggle and that we rally to our cause as many other nations of the free world as possible. The vast number of peoples on the side of communism, or at least under the dominion of the U.S.S.R., has made this struggle seem something of a race for population, but it is not so much a matter of numbers of people as of productivity. Productivity has come to be recognized as perhaps the most important economic concept of our time.

It is only by increasing output per manhour worked that we will be able to build effective defense against the aggressive powers of communism. And it is only by this means that we will be able to enlist the effective support of the peoples of the free world in this cause.

There is an immense difference between the material

standard of living in this country and that of the remainder of the world. Since 1945 the United States, under the Marshall Plan, has engaged in the largest exportation of national wealth to aid other peoples in the history of the world. This program has served a very important purpose. But the majority of this aid was used to rebuild Western Europe. The Middle and Far East, and other underdeveloped areas, whose standards of living are frighteningly below Europe, let alone our own, received only a small portion of this aid. Yet we cannot hope to repeat in these areas what we did under the Marshall Plan without severely reducing our own national wealth. What is needed is a means of very materially increasing the productivity of the peoples of these countries, as well as furthering the mechanization of European industry. Automation offers just such a possibility.

AUTOMATION IN UNDERDEVELOPED AREAS

Through the use of automation in the basic manufacturing industries, which are necessarily the first required in most underdeveloped areas, it may be possible to increase the material standard of living in these areas in a very short time. With automatic plants it is not necessary to train large labor forces. This fact is of immense importance. It is an enormous undertaking to train for the tasks of modern industry people who have made their living by agriculture, the tending of livestock, and the most simple crafts. The most prevalent idea about industrialization of underdeveloped areas is that it should be based on industries requiring large work forces. Is this really the most fruitful course?

If it is possible to introduce highly automatic basic manufacturing and processing plants in underdeveloped areas, to man these plants in part with a small nucleus of highly trained foreign technicians, and to supplement this group with a small trained work force from the local population, and thereby turn out a high per-capita rate of goods, might this not be preferable to a long-term, large-work-force industrial development plan requiring substantial changes in the patterns of life?

Automatic plants provide high industrial output without the need to alter the village and town structure of rural society, whereas the large-work-force industries necessitate changing this structure. The latter industries lead to concentration of population, requiring capital investment in housing, not to mention its questionable social desirability. The housing investment drains off much of the industrial investment that could be used instead to obtain industrial output.*

It is not at all certain that underdeveloped areas must undergo as long a process of industrialization as we have undergone and which most colonial development experts feel is necessary. Professor Wassily Leontief of Harvard has suggested the possibility that entire stages of skilled-labor development might be skipped by the introduction of automatic plants into underdeveloped areas, in the way the American chemical industry virtually skipped the generation of highly trained chemical technicians who participated in the development of the German chemical industry.

But as helpful as automation may be in increasing living standards of underdeveloped areas, it will nevertheless involve many problems. Can the long years of political development required before agrarian underdeveloped populations can become effective and respon-

* R. L. Meier of the University of Chicago has done some of the most progressive thinking and writing on this subject.

sible citizens of our modern society be by-passed? Who would control the automatized industry in these countries? How can even the small number of technicians needed by automatic plants be trained? Must the people of these areas adapt themselves to western products, or will they produce quite different products keyed to their own cultural patterns? How can the economies of these countries be kept from getting out of balance—more of one product than can be used without enough of another and perhaps complementary product? How can effective distribution be achieved?

If solutions can be worked out for these and other problems, automation may well provide a short cut to achieving a high rate of output of material goods among our allies and thus help to erect an effective block to further victories by the false god of communism.

WIDESPREAD EFFECTS

Since automation is an important new phase in a movement as basic to the history of man as mechanization, its effects are bound to be widespread. Indeed it is very difficult to consider any one economic or social effect without finding oneself exploring the effects of automation on all phases of our life and society.

For example, if we are to benefit from and encourage the use of automation in underdeveloped areas we must clearly trade with these areas. Yet our trade policy in recent years has been drifting and is therefore vulnerable to restrictions advocated by special interest groups. This state of affairs is likely to result in the erection of barriers to trade precisely at a time when it is necessary to adjust to the new patterns of resource and finished goods trade that the industrial changes of automation will bring.

Only a few of the most obvious social and economic implications of automation have been discussed in this chapter. But it is hoped that this discussion will lead to the asking of more useful questions than have thus far received "top billing" when the issue of automation has been raised. To say that automation is a worthy goal which will, on the whole, be beneficial is not to say that possession of that goal will not raise problems of even greater difficulty than were overcome in achieving it.

Where does the responsibility lie for training workers whose skills have been surpassed by machines? Will labor unions demand more government control of business and a more important role in politics as automation progresses? What changes will occur in our cities as industry moves to less populated areas? Just how much will automation affect plant location?

Although automation will result in more efficient use of our physical plant—for example, the use of capital equipment twenty-four hours rather than eight hours a day—our need for power and natural resources will continue to increase during the next generation. What steps must we take to replenish what we draw from nature?

What will be the effect upon business cycles of the increased capital requirements needed to build automatic plants? Must potential net profits in an age of automation be greater than they are today in order to justify assumption of the risk that a rapidly changing technology will make obsolete heavy commitments in plant and equipment? Does our present tax structure encourage the growth of automation? With virtually all labor "indirect" in an automatic plant, how will we allocate costs? Will a systematic basis for the analysis of maintenance functions be devised as this neglected branch of management grows in importance?

Increased knowledge of the communication process, which lies at the heart of the new technology, can lead to important changes in our approach to social organization. Will this mean that the way is open to concentration of power in the hands of a small group of people?

These are but a few of the questions which will be asked as the new technology grows in importance. Automation, however beneficial, will raise very real problems for the human race; just as all far-reaching social change, however beneficial, brings problems.

But, these problems are not altogether new. Just as automation is part of a longer continuum, so too the problems which automation will raise have been with us, in varying forms, for many years. Some of these problems seem to solve themselves, while others require a conscious effort for solution. Many, as is all too evident from the world about us, have not yet been solved. For it is indeed hard to provide a society in which increased material welfare truly benefits man rather than cheapens him. Strong moral leadership and men of good will are sorely needed, as much now as always.

Index